Master of Adaptation: How Bacillus subtilis Thrives Everywhere

Jacksin

Contents

1 Literature review

1.1 *Bacillus subtilis* as a host organism

Bacillus subtilis is a Gram-positive, spore-forming, rod shaped bacterium with low % G+C in its genome capable of growing in diverse environments [43, 31]. Some of its most common ecological niches include soil, plant roots and the gastrointenstinal tracts of animals [173, 5, 186]. Formerly it was believed that *B. subtilis* is obligatory aerobe however experimental evidence proved that this bacterium is capable of growing anaerobically as well, by utilizing nitrate instead of oxygen as an electron acceptor [132].

The very first application of *B. subtilis* was for the production of natto, a Japanese food product consisting of fermented soybeans. Since then it became subject of extensive studies. The announcement of the genome sequence of *B. subtilis* 168 facilitated one of the most powerful tools that broadened the spectrum of knowledge and aided in the genetic transformation of this bacterium for industrial settings [102]. Nowadays *B. subtilis* is the best characterized Gram-positive bacterial in regard to genetics, biochemistry and physiology [174, 191, 161].

What made *B. subtilis* one of the most popular bacteria to study, is its extraordinary ability to follow different developmental programs such as natural competence, biofilm formation, cell motility and sporulation [115]. From the industrial point of view, *B. subtilis* fulfilled many criteria that qualified it to evolve in a powerful industrial workhorse for the overproduction of heterologous proteins and fine biochemicals. Some of the main reasons that made *B. subtilis* stand out are i) its high genetic amenability stemming from its natural competence for uptake of DNA ii) its excellent fermentation properties and iii) its ability to secrete proteins directly to the medium [34, 31, 46]. These characteristics favor its selection over other commonly used hosts such as *E. coli* where the produced proteins usually accumulate in the cytosol thus leading to the formation of inclusion bodies [47, 177].

Today owing to the advances in synthetic biology and the development of new strategies for metabolic engineering, the application portfolio of *B. subtilis* has significantly expanded from production of recombinant proteins to pharmaceuticals, vitamins and other bio-products [188, 119]. To this end, optimized chassis have been constructed such as proteases deficient strains for increased production of proteins. In the present work the seven proteases deficient strain *B. subtilis* KO7 which is derivative of the laboratory host PY79, has been employed (*Bacillus Genetic Stock, Center ID 1A1133*). Besides its role as a cell-factory, *B. subtilis* spores have been extensively used as probiotics in aquaculture and agriculture while by developing the spore surface display technology, spores have been used as carriers of industrial enzymes [34, 46]. Recently *B. subtilis* biofilms and spores have been employed for the production of biomaterials [63, 74]. The applications of spores will be discussed in detail in the sections to follow.

1.1.1   Sporulation in *B. subtilis*

In the 19th century Ferdinand Cohn [28] and Robert Koch [93] independently noticed that certain bacteria were able to switch from vegetative state to dormancy by developing a resilient cell form called endospore. Endospore (or spore) formation in bacteria is a survival mechanism triggered mainly by starvation signals. Sporulation is energy and time consuming process that after 2 hours of initiation becomes irreversible [42]. Thus commitment to this cellular process is a so called "last resort" response to starvation that the cell undergoes once all the other stress coping mechanisms have failed [155]. The purpose of sporulation is to generate a metabolic inactive cell that will maintain the genetic information under various environmental insults including UV and gamma radiation, wet and dry heat, desiccation and oxidizing agents, enzymatic lysis and noxious chemicals [138]. Despite their inert state spores retain the ability to detect the nutrient availability of the surrounding environment and revive when the growth conditions are restored undergoing germination. Thanks to its architecture, a dormant spore can survive from decades to thousands of years [89, 180]. This remarkable resilience is associated with many features such as the partially dehydrated core and the protective multilayered coat [35].

Among a variety of spore forming bacteria, more in depth research has been conducted for the genera of *Clostridia* and *Bacillus* with *B. subtilis* being one of the best described examples. Sporulation in *B. subtilis* (Figure 1.1) follows a peculiar cellular mechanism which initiates with an asymmetric cell division generating a sporangium comprising of two unequal cells, each containing a single copy of the chromosome [187, 45]. The large cell is called "mother cell" and is responsible for the formation of the forespore which takes place in the small compartment. The two compartments stay in contact through a peptidoglycan wall called the "polar septum" [187, 45]. Next, the polar septum bends and the mother cell engulfs the forespore which gradually will mature in a resilient spore [187, 45]. When formation is completed, the mother cell undergoes lysis in order for the spore to be released [187, 45].

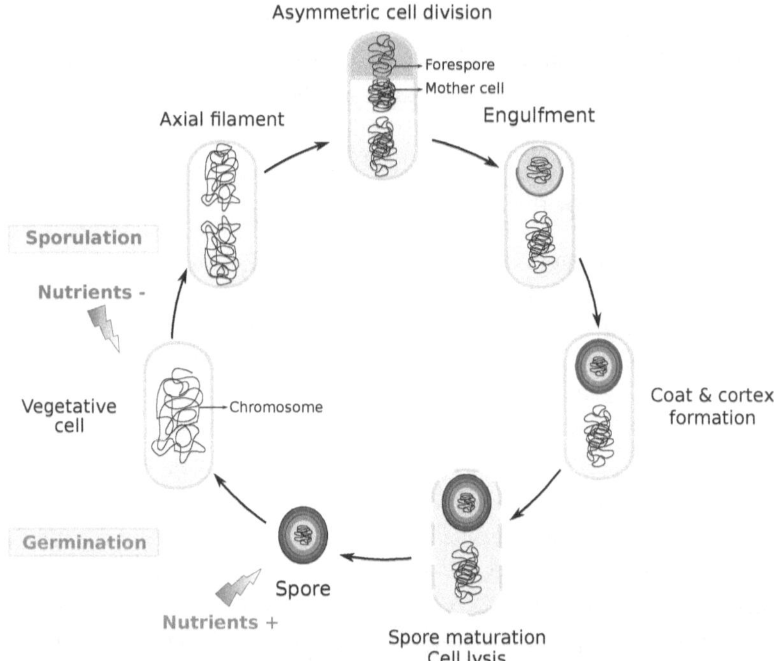

Figure 1.1: Schematic representation of the key stages of sporulation cycle in *B. subtilis* adapted from [124]. Sporulation initiates from nutrient starvation signals that lead to asymmetric cell division. Following the forespore compartment is engulfed by the mother cell. Upon engulfment spore coat formation initiates. Finally the spore matures within the mother cell which lyses in order to release the spore. Upon nutrient availability the spore can revive following a process called germination.

1.1.1.1 Overview of sporulation regulation

The starvation signal triggers a cascade of molecular reactions which involves the finely tuned regulation of over 500 genes in the course of approximately 8 hours [130]. One of the critical steps for entry to sporulation is the activation of the transcriptional factor Spo0A via phosphorylation (Spo0A~P) [56]. However since sporulation is an energy consuming and irreversible developmental mechanism, several alternative regulatory pathways are activated in order to delay or even prevent initiation of this process. Therefore Spo0A is subject to negative regulation as well such as dephosphorylation from the phosphatase Spo0E [144]. When enough Spo0A~P has accumulated then different subsets of genes are activated which are required for axial filament formation and asymmetric division. Furthermore accumulation of Spo0A~P sets in motion activation of compartment specific sigma factors [133].

Sporulation commences after *B. subtilis* chromosome has replicated one last time [194]. The two identical chromosomes are anchored with the origin of replication to the opposite cell poles, forming an elongated structure called "axial filament" [189]. Following, the cell divides asymmetrically by forming a septum near the one pole, trapping 30 % of the chromosome in the forespore compartment and causing a temporary genetic asymmetry since only the genes included in this part of the chromosome can be expressed [153]. The remaining 70 % of the chromosome is actively transported in the forespore by the membrane translocase SpoIIIE [153, 114]. Formation of the septum is one of the key stages in sporulation which sets in motion distinct gene expression programs in the two compartments [45]. The genetic differentiation is linked to the activation of compartment specific sigma factors i.e. σ^F and σ^G in the forespore and σ^E and σ^K in the mother cell [51].

After asymmetric division the forespore specific σ^F is activated. Prior to activation σ^F is kept inactive by the anti-sigma factor SpoIIAB. In a similar way to σ^F, σ^E is produced already prior to asymmetric division in both compartments but it is held in an inactive pro-σ^E form [187]. After septation σ^E is activated only in the mother cell by the SpoIIGA protein [187].

After asymmetric division and activation of σ^F and σ^E, follows prespore engulfment. During this event the septum cell wall is remodeled and the mother cell membranes encase the forespore compartment [66, 187, 45]. At the last step of engulfment the spore detaches from the cell wall and starts shaping the spore cortex which participates in the dehydration of the spore. Inter cellular communication between the two distinct cells is achieved through a channel consisting of the proteins SpoIIAA-SpoIIIAH and SpoIIQ [16, 187]. Through this channel the mother cell is able to provide with nutrients the prespore thus supporting gene expression in this compartment [16]. Through this communication channel σ^F activates the last prespore specific sigma factor, σ^G [66, 187, 45]. The last mother cell specific sigma factor σ^K is first produced in an inactive form which undergoes posttranslational modifications in order to be transformed in the active form [101]. σ^K along with the small DNA binding protein GerE regulates the formation of most of the coat proteins and in collaboration with σ^G the synthesis of the cortex [39].

1.1.2 The spore composition and morphology

1.1.2.1 The cortex

The mature spore contains the genome in the partially dehydrated core where Ca^{2+}-dipicolinic acid (CaDPA) is mostly present. The DNA gets saturated by α/β-type small, acid-soluble spore proteins (SASPs) that alter its conformation ("A-B-DNA") thus protecting it from UV irradiation and photodamage [105]. The low water content (25-50 % of wet weight) of the core and the binding of α/β-type SASP on the DNA, are mainly responsible for the extraordinary spore resistance [166]. Besides DNA, the core contains RNA and enzymes that are important for germination and outgrowth [98, 27]. Due to the low water content of the core these enzymes remain inactive until germination. Until this point, spores remain dormant and have no metabolic activity. Since spores may have to remain to this state for very long time, the core contains no to little amounts of high energy molecules such as ATP and NADH [165].

The core is surrounded by inner spore membrane, a thin layer of germ cell wall, cortex, outer spore membrane and at outermost from a multilayered coat (Figure 1.2A)[125, 45]. Synthesis of these layers begins during the engulfment stage of sporulation. The cortex is composed of modified peptidoglycan (PG) localized between the inner and outer membrane layers encasing the spore core. The cortex differs in structure from the vegetative cell wall due to the presence of muramic δ-lactam residues and lower frequency of transpeptidation and cross linking of the glycan strands [149, 187]. Although its composition differs to the cell wall PG, its synthesis shows many similarities [149]. The cortex together with the underlying germ cell wall which is composed of PG similar to that of the cell, maintain the integrity of the spore inner membrane [166].

1.1.2.2 The coat

The spore coat is made up of more than 80 proteins distributed over four layers: a basement layer, the inner coat, the outer coat and the most outer shell called "the crust" (Figure 1.2) [187]. The coat mainly consists of proteins while polysaccharides represent a smaller fraction [39, 172]. Cross-linking of the spore coat proteins confers to the spore rigidity and mechanical stability [209]. However the spore coat is not impermeable, since germinants can pass through and reach the germination receptors located in the inner membrane thus triggering germination [40]. Furthermore, many enzymes with important role in germination such as spore cortex lytic enzymes CwlJ, SleB and GerQ are located on the spore coat [97, 3]. The structure and the size of the coat can differ depending on the sporulation conditions (e.g temperature, cations concentration) [1, 81].

Formation of the spore coat is a dynamic process that starts prior to engulfment. Assembly requires correct protein localization and spore encasement. Coat protein localization initiates at the mother cell proximal (MCP) pole forming a scaffold cap. With the scaffold cap being the starting point, coat proteins expressed in three waves encase the spore [125, 124]. Most, if not all, of the spore coat proteins are produced in the mother cell compartment [209]. Previous studies have shown that the spore coat proteins follow an uneven distribution i.e. certain proteins are more concentrated on the spore poles rather than the middle of the spore [78]. The proper assembly of each spore coat layer fully depends on expression and correct synthesis of key regulatory proteins (SpoIVA, SpoVID, SafA, CotE and CotXYZ respectively) called morphogenetic proteins [187, 125, 45]. The role of the morphogenetic proteins is to interact with a certain group of coat proteins and target them on the corresponding coat layer [125]. However to date still the exact interaction mechanism between the morphogenetic proteins and the coat protein remains unclear. A simplified form of this interaction network is presented in Figure 1.2B.

Basement layer: The assembly of the basement layer depends on the production and interaction of three proteins: SpoIVA, SpVID and SpoVM. SpoIVA localizes on the mother cell side of the forespore and assists in the assembly of the basement layer with the help of SpoVM [187, 40]. The basement layer serves as

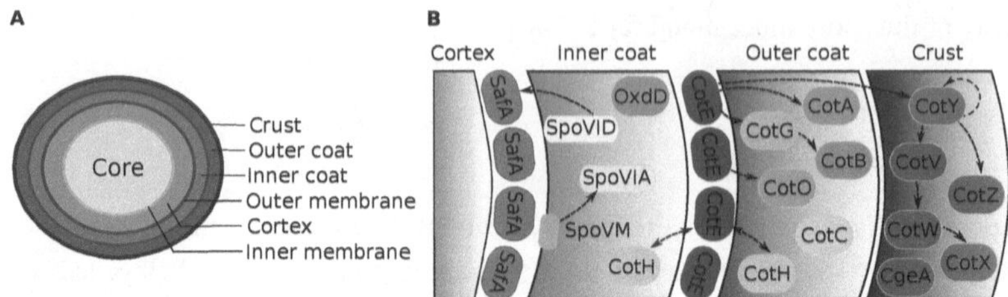

Figure 1.2: Composition of the *B. subtilis* spore coat adapted from [124]. **A:** Representation of the spore. **B:** Simplified representation of the protein interaction network regulating formation of the spore coat layers.

scaffold for the deposition of the other layers. In *spoIVA* mutants, the spore coat is synthesized but is not attached on the forespore. Furthermore *spoIVA* mutants are not able to form the cortex [41, 40]. Similar to SpoIVA, SpoVID protein is involved in the anchoring of the spore coat around the forespore but its synthesis is not crucial for the formation of the cortex [40].

Inner and outer coat: Assembly of the inner coat depends on SafA which localizes in the cortex/coat interface [141]. Localization of SafA depends on protein interactions with SpoVID [141]. Electron microscopy analysis on *safA* mutants revealed that the inner coat layer is absent from the mature spore [185, 124]. Crosslinking of SafA monomers is conducted by a transglutamase called Tgl [49]. One of the components of the inner coat is the protein OxdD which encodes for oxalate decarboxylase [30].

Formation of the outer coat is under the control of CotE. *cotE* mutants lack the outer coat and the crust [207, 124]. Barák et al. conducted a large scale screening elucidating the protein interaction network between CotE and the related coat proteins [100]. Although a large sub group of protein require CotE for correct assembly on the spore coat, only a few were found to directly interact [100]. Finally for the correct assembly of the respective layers, coordination of the morphogenetic proteins in protein level is pivotal, i.e. recruitment of SafA and CotE depends on SpoIVA whereas recruitment of CotXYZ depends on CotE [40].

Crust: One of the most recent discoveries regarding the morphology of the *B. subtilis* spore, was the existence of an additional layer located in the most outer part of the spore. This layer is called "crust" and is identified as the "insoluble fraction" of the spore coat due to the abundance of cystein rich proteins which get crosslinked [203]. At least six proteins have been involved in its formation: CotVWXYZ and CgeA. According to genetic interaction studies, recruitment of CotXYZ depends on CotE [40]. McKenney et al. proposed CotXYZ as the morphogenetic proteins of the crust [126]. *cotXYZ* null mutants fail to localize the CotW protein and thus have deformed shapes [126]. Furthermore these mutants seem to produce more hydrophobic spores compared to will type, suggesting the putative role of the crust in the regulation of the spores hydrophilicity [171].

Analysis of the spores employing ruthinium red staining and electron microscopy (EM) in combination to other evidence suggest the existence of glycosylation sites on the crust [195, 17]. Screening for glycosylation motifs in the amino acid sequence of the crust proteins suggested CotX as a candidate for glycosylation [8]. Furthermore glycosylation of the spores was associated with the *sps* genes. Mutants lacking these genes demonstrate higher hydrophobicity while they seem to be more prone to formation of spore clumps [17].

1.1.2.3 Coat proteins employed in the present book

In the present doctoral book in total four anchoring proteins (CotB, CotG, CotY, CotZ) were employed for spore surface display. CotB and CotG contribute to the formation of the outer coat while CotY and CotZ are components of the crust.

CotB and CotG are two abundant outer spore coat proteins. The genes for *cotB* and *cotG* were identified by reverse genetics [37]. Both of them are transcribed at the late stage of sporulation under the control of sigma factor σ^K and GerE. Their assembly on the outer spore coat is governed by CotE. *cotB* encodes for a 46-kDa protein but due to posttranslational modifications its size is extended to 66-kDa [208]. Fusion of GFP to CotB has revealed its localization at the middle of the spore as a ring structure [78]. For the correct formation and deposition of CotB on the outer spore coat direct interaction with CotG in the presence of CotH is necessary [208]. Thus spore mutants that lack *cotH* or *cotG* are not able to form CotB. However loss of CotB has no apparent effect on the spore resistance nor on spore coat structure [146]. *cotG* encodes for a 24-kDa protein containing nine tandem repeats of a 13 amino acid long sequence in its central part [159]. Unlike *cotB* mutants, strains lacking *cotG* form an amorphous outer coat thus underlying its role in determining the structure of the outer layer [54].

CotY and CotZ are localized in the crust and they have a pivotal role in the coordination of the assembly of this layer. CotY is the most abundant crust protein followed by its paralogue, CotZ [99]. The last one besides being one of the main structural proteins of the crust has another really important role as morphogenetic protein responsible for anchoring the crust in the middle of the spore. CotY and CotZ (18- and 16-kDa respectively) are cystein rich proteins (representing ~10% of their protein sequence) that share peptide motifs. Due to the high cystein content and the disulfide bonds that are formed between the residues, these proteins form stable structures that are highly insoluble [99]. Thus heterologous expression of CotY and CotZ in *E. coli* or even their extraction from the spore coat is a great challenge [99]. Research on how CotY assembles revealed that this protein can form hexameric proteins that can further be arranged in sheets [82].

1.1.3 Properties of the spore

Spores exhibit great resistance against a broad range of assaults including for instance: UV and γ-radiation, enzymatic degradation, chemicals, oxidizing reagents, high pressure, desiccation, extreme temperatures etc. A variety of factors contribute to the resistance of the spores including the spore coat and the enzymes that protect the DNA.

The spore coat serves as a first line barrier that protects the spore against environmental assaults. The coat is a multilayered construct consisting of highly crosslinked proteins. This compact structure retains the rigidity of the spore against mechanical stresses [83]. Furthermore the coat restricts access to big molecules like lysozyme thus making the spore less prone to predation [91]. Another characteristic of the spores, is their resistance to toxic chemicals that are lethal for growing cells. This property is the consequence of many factors including detoxifying enzymes located on the spore coat such as catalases (KatA, KatB, KatX) to decompose hydrogen peroxide and the superoxide dismutase (SodA) to eliminate superoxide [157]. One major requirement for spores defense against noxious chemicals is the integrity of the spore coat. Spores lacking the morphogenetic protein CotE are not able to form the outer coat thus are more sensitive to chemical assaults [207]. Furthermore a key factor to spores resistance is the low permeability of the inner membrane to small molecules such as water. According to analysis of the lipid composition and structure of the inner mebrane, albeit the similarity of its composition to the membrane of growing cells, the lipids are largely immobile [165].

Chemicals are not only capable of damaging the spore structure but also can be fatal by causing DNA damages. Such a genotoxic chemical for instance is formaldehyde. Spores protect their DNA by saturating it with α/β-type SASP. SASPs mainly bind in DNA's minor groove thus sensitive groups located in the major groove are exposed. Hence conformational change of the DNA due to biding of SASPs protects the DNA against photodamage. Radiation of spores with UV generates a low amount of photoproducts in contrast to radiation of growing cells where the impact is much higher [168]. Thus α^{-}/β^{-} spores are more sensitive to chemicals and UV radiation [165, 168]. Many of the DNA damages acquired during sporulation can be elevated through DNA repair mechanisms during outgrowth.

Besides the major role of the α/β-type SASP, the spore core low water content and the high concentration of CaDPA render in the protection of the DNA against radiation and heat. For instance, low water content reduces the mobility of the proteins stored in the core, thus protecting them against thermal inactivation and aggregation [168].

1.1.4 Applications of the spores

Spores have been extensively studied due to their extraordinary properties. A significant part of research has focused on investigating the mechanisms that underlie these properties while another part has been dedicated on their biotechnological applications. Spores due to the architecture of the coat, their inert nature and their resilience combined with the low cost production, are ubiquitous biomaterials [205].

One of the most common uses of the spores has been as probiotics. In humans probiotics have a widespread use such as dietary supplements or prophylactics (for instance to prevent childhood diarrhea) [70]. In aquaculture probiotics are employed to promote growth and disease resistance in various fish and shrimp [142, 134]. More recently spores have been used as building blocks for the production of living materials. González et al. embedded spores into an agarose hydrogel in order to form a robust biomaterial [63]. Thanks to their ability to withstand high temperatures, spores could survive the conditions of 3D printing while they were able to be stored long term after desiccation. Once the gel was re-hydrated with nutrients, spores initiated germination. Therefore the authors utilized this natural ability of the spores, to design biosensors and to produce antimicrobial peptides [63]. Following a similar concept, spores were utilized for the construction of self-healing concrete. Sarkar et al. used a modified *B. subtilis* strain that was able to express a biosilification peptide [160]. Cracked mortars containing spores were incubated in media which induced germination and outgrowth. After 28 days the cracked mortars were healed to a great extend, thus indicating the expression of the biosilification peptide from the vegetative cells and the incorporation of gehlenite [160]. Spores can absorb heavy metals on their surface thus have been used for bioremedation [73]. Due to this natural ability of the spores, a few studies suggest their employment for the extraction of rear earth ions. Dong et al. demonstrated *B. subtilis* spores could bound Tb^{3+} and Dy^{3+} in the most outer coat [36]. Finally one of the most widespread applications of the spores has been as platform for immobilization of industrially relevant enzymes of antigens. This particular application of the spores which is relevant for the present book will be elaborated in detail in the following section.

1.2 *In vivo* protein display

Biological units such as cells, naturally present on their surface a variety of proteins that assist for example in intercellular communication, signal transduction, surface adherence and motility. Scientists developed the protein presentation technology by fusing proteins or peptides of interest (POI) to naturally exposed proteins. The most common *in vivo* approaches for protein display involve yeast or prokaryotic cells, phages and bacterial spores [58]. One of the main advantages of employing biological platforms for protein display is the coupling of genotype and phenotype. Therefore protein display is widely used for screening peptide libraries and enzyme variants produced by directed evolution [50]. Besides that, surface display can be employed for the development of live vaccines, biocatalysts by immobilizing enzymes, biosensors by exposing signal sensitive molecules such as receptors or even sorbents [107].

The concept of protein display was introduced by George Smith in 1985 who generated recombinant filamentous phages by fusing antigens to the coat protein pIII of the virion [178]. Since then, phage display became one of the most widely used tools for creating antibody libraries [200]. Selection of the desired antibodies or peptides is done through biopanning process and highly depends on the binding affinity of the protein to the target molecule [200]. The later one is often immobilized on a surface. The selected phages are then isolated and enriched through bacteria infection [143]. Although this method is ideal for high through-put screening and enables generation of vast protein libraries, it is hampered by the non-trivial protocols and the large number of false positives. Furthermore, since the POIs are synthesized in the bacterial host, the reducing environment of the cytoplasm may affect their proper folding [143].

In the case of bacteria cell surface display the first reports were published in 1986 by Freudl et al. and Charbit et al. [55, 21]. Using cells as carriers has the ad-vantage of auto-replication thus circumventing the need for capture of the carrier and subsequent host infection as happens with phages. Furthermore bacteria cell size is compatible with high throughput methods such as fluorescent activated cell sorting (FACS) which makes it ideal for screening large enzyme libraries for activity [84]. Becker et al. combined *E. coli* surface display and FACS for high throughput screening of esterases with enhanced enantioselectivity against flu-orescently labeled substrates [9]. One of the most frequently used bacteria for protein display is the Gram-negative organism *Escherichia coli*. The popularity of *E. coli* derives from the genetic amenability and the high transformation effi-ciency. Gram-negative cells commonly contain three types of membranes: the in-ner membrane that typically consists of phospholipids, the peptidoglycan cell wall and the outer membrane which is composed of phospholipids and lipopolysac-charides [176]. Therefore the protein to be displayed first must cross the inner membrane, the periplasmic space and last the outer membrane. Consequently proper secretion of the POI needs to be secured. The most frequently used carrier protein in *E. coli* is a chimeric protein composed of a region of the lipoprotein Lpp and the transmembrane segment of the outer membrane protein A (OmpA) that can expose proteins on the outer side of the cell [84, 183]. The protein of interest is attached to the C-terminal of the later one [84, 183]. This system has been broadly utilized for the display of enzymes, single-chain variable fragment (ScFv) and antibody fragments. However its applicability is hampered by its low effi-ciency to display proteins with extensive secondary and tertiary structures [199]. An alternative choice for protein display in Gram-negative bacteria, is the use of the filament proteins [26]. Such an example is the flagellin protein FliC and the protein of the flagellar cap FliD which has been used for the display of small peptides [121]. A more recent display system includes the utilization of auto-transporter proteins and ice nucleation proteins as carriers [206]. These proteins contain a mobile β-barrel anchor thus allowing the displayed monomeric proteins to move across the outer membrane and form active multimeric proteins [179].

The main limitations of bacteria surface display are the inability to perform complex posttranslation modifications, the expression bias for mammalian proteins and the low cell tolerance in toxic byproducts [179].

Yeast surface display was first developed by Boder and Wittrup in 1977 [12]. Since then it has been used for multiple purposes including protein and peptide library screening, epitope mapping, protein engineering, whole cell biocatalysis [96, 112]. The most frequently used organism for yeast display is *Saccharomyces cerevisiae*. Throughout the years different receptors have been tested for the display but the most commonly used is the Aga1-Aga2 mating proteins [112]. In this system, the POI is fused to the C- or N-terminal of the Aga2 which is linked to the Aga1 subunit of the mating protein located on the yeast surface. The POI is directed to the extracellular space by signal peptides that transport it to the outer membrane. In order to enable quantification of the expression levels, most commonly the POI is flanked by epitope tags (e.g. c-Myc and hemagglutinin antigen (HA) tags) allowing for antibody labeling and subsequent quantification by flow cytometry analysis [110]. Lim et al. demonstrated that Aga2 could be used in a dual fusion thus simplifying quantification of binding interactions by fusing one side of the Aga2 to a fluorescent protein [110]. The popularity of yeast surface display relies on many factors. First, due to the eukaryotic expression machinery proteins can be modified posttranslationally (e.g. disulfide bonds) [57]. Next one major advantage is the ease to perform genetic modifications. Finally, due to size of the cells analysis can be easily performed with flow cytometry combined with cell sorting [57]. Thus it's suitable for high throughput screening. Usage of this system is hampered by the low transformation efficiency of yeasts, the slow cell growth and from cell viability issues under extreme conditions [179].

1.2.1   *B. subtilis* spore surface display

Spore display is a protein immobilization method that allows the presentation of peptides or proteins on the spore surface. Two methodologies have been reported so far for display; a recombinant and a non-recombinant. The former strategy takes place *in vivo* and most frequently entails the genetic fusion of the gene of interest (GOI) to one of the genes encoding for spore coat proteins thus ensuring the deposition of the chimeric protein on the spore surface during sporulation [23]. Recently Chen et al. reported an alternative to this strategy, in which the proteins of the coat were genetically fused to cellulosomic cohesin domain while the POI was genetically fused to a docerin domain [25]. The two genetic constructs were co-expressed during sporulation while immobilization was governed by cohesin-docerin interactions with the mother cell facilitating the deposition on the spore surface [25] In the non-recombinant approach, immobilization of the POI on the spore surface is conducted by physical adsorption or chemical immobilization [59, 48]. In this case, most frequently the spores and the POI are separately produced while immobilization is conducted in a following step [48]. In the current book spore display was carried out following the recombinant approach by fusing the POI to the spore coat proteins.

The first spore display was performed by Isticato et el. in 2001 to present the fragment of the tetanus toxin (TTFC) on the *B. subtilis* spore surface as a fusion to the coat protein CotB [80]. Since then various applications of spore display have emerged including vaccine development, directed evolution, bioremedation and biocatalysis [111, 205]. Among the spore forming bacteria the most commonly used organism for spore display is *B. subtilis* due to its genetic amenability and the available knowledge on genetic, biochemical and physiological level regarding spore formation [174, 191, 161].

The spore is a metabolically inert cell unable to divide that consists of a dehydrated core surrounded by four different layers [187]. Major component of these layers are proteins that secure spore's rigidity and resilience against plethora of environmental assaults such as extreme temperatures and pH, broad range of chemicals etc. More about the properties of the spores are described in Section 1.1.3. Spore formation takes place within the sporulating cell with most of the components of the spore coat being synthesized in the cytoplasm and then directed from the mother cell on the developing spore. When spore formation is completed then the cell lyses and the spore is relished [187].

A simplified representation of the design of fusion genes for spore surface display is shown in Figure 1.3. The spore coat proteins can be linked to the POI either C- or N-terminally. In many cases a linker is introduced between the two sequences to prevent steric hindrance of the displayed POI [135, 67]. The genetic construct can be integrated in the chromosome of *B. subtilis* via homologous recombination or it can also be expressed in the form of replicative plasmid [87]. Transcription of the fusion gene can be regulated by sporulation specific promoters such as the native promoters of the coat proteins or by chemically inducible promoters such as P_{grac} [136]. After synthesis of the fusion protein, deposition on the spore coat

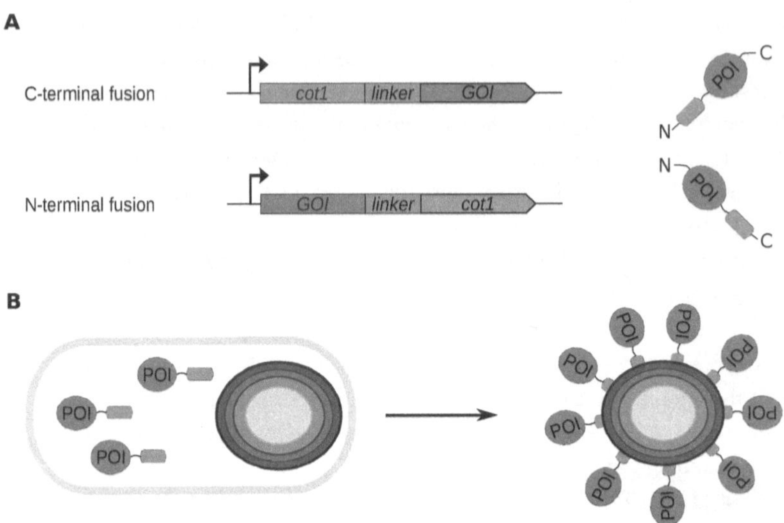

Figure 1.3: Schematic representation of the protein display technology on *B. subtilis* spores. **A:** The gene of interest (GOI) is genetically fused to the gene of the coat protein (*cot1*). Arrow represents the sporulation inducible promoter. The synthesized peptide/protein of interest (POI) is fused to the C- or N-terminus of the coat protein (green). **B:** Cellular process of displaying fusion proteins on the surface of the spore during sporulation.

is regulated by the mother cell compartment during sporulation. Many spore coat proteins have been tested for spore surface display (Table 1.1). Most frequently spore coat proteins of the outer coat layer are employed such as CotB, CotG and CotC [111]. More recently studies reported on the utilization of the proteins forming the crust such as CotY, CotZ and CotX for display [111].

Table 1.1: List of coat proteins and proteins/peptides of interest (POI) used for *B. subtilis* spore display and the related applications.

Coat protein	POI	Application	Ref.
CotB	Thermophilic lipase (Tm1350)	Biocatalysis	[23]
-	Thermostable esterase	Biocatalysis	[22]
-	VP28	Vaccine development	[139]

Table 1.1: List of coat proteins and proteins/peptides of interest (POI) used for *B. subtilis* spore display and the related applications.

Coat protein	POI	Application	Ref.
-	Tetanus toxin C (TTFC)	Vaccine development	[80]
-	18 x Histidine	Bioremedation	[68]
CotB, CotG, CotC	Urease subunit from *Helicobacter acinonychis* (UreA)	Vaccine development	[69]
CotB, CotG	Flagellar cap protein from *Clostridium defficile* (FliD)	Vaccine development	[150]
CotG, OxdD	Phytase	Animal feed additive	[129, 151]
CotG	Haloalkane dehalogenase (DhaA)	Biocatalysis	[196]
-	Nitrilase	Biocatalysis	[24]
-	*Lactobacillus rhamnosus* p75 protein	Probiotic	[85]
-	Chitinase	Biopesticide	[158]
CotG, OxdD, CotY, CotZ	β-galactosidase	Biocatalysis	[25, 151, 197]
CotE	Tyrosinase	Bioremedation, biosensor, biocatalysis	[71]
CotC	Surface immunogenic protein from *Streptococcus agalactiae* (Sip)	Vaccine development	[202]
CotC, CotG	Trehalose synthase	Food additives, cosmetics, pharmaceuticals, agricultural products	[113]
CotV, CotW, CotY, CotZ, CotX, CgeA	Laccase	Biocatalysis/Directed evolution	[7, 64]

1.2.1.1 Advantages and limitations of spore display

Using spore display has many advantages over the traditional cell surface display strategies. To begin with, protein synthesis of the fusion protein and immobilization on the spore coat take place intracellularly. In this process no membranes are present, hence the fusion protein is not translocated as happens in cell surface display. For this reason spores can be employed for display of large and multimeric proteins such as β-galactosidase [25, 151, 197]. Correct folding of the protein is ensured by the presence of chaperones that are naturally expressed during sporulation [90]. Importantly, spores retain a resilient structure that enables them to survive under harsh conditions where most cells can not [92]. Therefore spores are prominent immobilization platforms for biocatalysts.

From an industrial perspective, production of recombinant spores in large scale is feasible since Bacilli spores are already produced in bulk amounts and are used as biopesticides [10]. The industrial applicability of the spore display system is further highlighted by the reduced production costs. In particular, spore display facilitates the concurrent enzyme production and immobilization on the spores, while purification is conducted by a few centrifugations. Overall the number of steps required from production to the final biocatalytic process are less compared to the in *vitro* enzyme immobilization strategies where the carrier and the biocatalyst are produced by separate processes [156]. Additionally production of the spore-biocatalyst is an eco-friendly process since it avoids the utilization of toxic substances.

In contrast, one of the main limitations of the spore display technology is the lack of tools for quantification of the displayed protein. For the *in vitro* immobilization methods the protein immobilization yield can be estimated by measuring the amount of protein in the solution before and after immobilization [4, 59]. However in spore display this is not possible since the immobilization takes place in *vivo*. Furthermore the spore coat is a really robust structure that can not be easily extracted by chemical or physical approaches. As a result only a certain amount of protein can be extracted from the spore coat. Furthermore, the spore coat consists of multiple enzymes which might be the source of undesired side reactions [64, 122]. Finally, the small spore size (typically around 1 x 0.5 μm) could be a limiting factor to the number of loaded protein molecules compared to display approaches that rely on cells which have a much larger size [18].

1.3 Context and aims of the book

Biotechnology in the broadest sense is the use of biological systems for the benefit of humankind. Its applications are categorized in five main sectors including agriculture (green), aquaculture (blue), medicine (red) and industry (white) [11]. White or industrial biotechnology employs enzymes and microorganisms for the production and processing of chemicals, energy and materials. Mankind has been relying on white biotechnology for thousands of years to develop goods such as cheese, bread and wine [11]. However the increasing population growth followed by higher energy and food demands, the depletion of petroleum-based fuels and the degradation of the environment, have opened new routes for industrial biotechnology. Nowadays, a global effort is in motion to develop bio-based strategies towards a sustainable economy [6].

Owing to the advances in synthetic biology and protein engineering, microorganisms and enzymes are being modified to replace resource-intensive processes while at the same time reducing the environmental footprint. Consequently biocatalysis has emerged as a promising technology for chemical manufacture. However the lack of long-term operational stability, the difficulties in retrieving and reusing the catalyst as well as the costs of production are the main limitations that make some biocatalytic processes industrially incompatible [169]. To this end, enzyme immobilization technology has evolved to be a key solution. The implication of multiple purification steps for the production of the enzymes, the materials used as carriers and the excessive protocols for immobilization increase the overall costs of the process several orders of magnitude [193]. Hence developing immobilized biocatalysts with lower production costs is of paramount importance.

The central objective of the present doctoral book was the development of a biological platform for immobilization of enzymes based on spores from *B. subtilis*. Envisioning the utilization of spores in industrial processes, the initial focus was set on the development of optimized strains that produce spores with significantly reduced germination rate and have rather high sporulation efficiency. Next, realizing the lack of available tools for quantitative assessment of non germinating spores, a flow cytometric method was developed that enables their separation and subsequent quantification. Following, the book aimed at the utilization of spores as carriers for immobilization of enzymes, using the fusion protein tech-nology. Within this framework, a library of coat proteins and linkers was designed and evaluated using fluorescent proteins as reporters. Taking into consideration these results two different spore display systems were designed for the production of biofuels and glycosides respectively. More precisely, a light activated fatty acid decarboxylase (*Cv*FAP) and two variants of sucrose phosphorylase (*Lm*SucP, *Ba*SucP) where presented on the surface of the spores and were tested in biocatalytic processes.

2 Common Materials & Methods

2.1 Materials

2.1.1 List of devices

Table 2.1: List of employed devices.

Device	Model	Supplier
Centrifuges	Micro Star 17 (Eppendorf Tubes)	VWR, Germany
	Micro Star 17R (Cooling Eppendorf Tubes)	VWR, Germany
	Mega Star 3.0R (Falcon tubes 15 & 50 mL)	VWR, Germany
Clean Bench	-	BDK
Cuvette photometer	Cadas 50S spectral photometer	Dr. Lange
Gel documentation	Camera CyberShot DSC-OX10	Sony
Gel electrophoresis	PB-0	Febikon
Electroporation chamber	MicroPulser	Bio-Rad Laboratories, USA
Fluorescence activated cell sorter (FACS)	LE-SH800SZBCPL	Sony
Gas chromatograph	Nexis GC-2030 capillary Column SH-Rxi 5MS	Shimadzu, Germany
High performance liquid chromatography (HPLC)	Agilent 1260 infinity bio-inert	Agilent Technologies
Incubation shaker	Multitron II	Infors, Germany
	Innova 44	New Brunswick, USA
Incubator (Agar plates)	Incu-Line Tower	VWR, Germany
Light microscope	AXIO Vert.A1	Carl Zeiss AG, Germany
Mixer mill	MM 400	Retsch GmbH, Germany

Table 2.1: List of employed devices (continued).

Device	Model	Supplier
NanoDrop	UV5Nano	Mettler Toledo, Germany
Thermocycler	DNA Tetrad 2 Peltier Thermal Cycler	Bio-Rad Laboratories, USA
Heat incubator Eppendorf	Thermomixer comfort	Eppendorf AG, Germany
UV table	Reprostar 3	CAMAG
pH-meter	HI 2211 pH/ORP Meter	HANNA Instruments Germany GmbH
Pipettes	Research Plus	Eppendorf AG, Germany
Power supply for electrophoresis	EPS 301	Amersham Biosciences Europe GmbH, Germany
Plate reader (UV/Vis-Spectrometer)	Clariostar	BMG Labtech, Germany
12 V LED Strips	LEDs emitting at 465-470 nm (LT3528B6050012)	Mextronic, Germany
Vortex	Reax Top	Heidolph, Germany
Robotic platform custom made (Analytik Jena, Germany)		
Robotic arm	PreciseFlex PF750	Precise Automation, USA
Pipetting Robot	CyBio FeliX Head R96	Analytik Jena, Germany
Incubator	Cytomat 2	Thermo Fisher Scientific, Germany
Plate reader (UV/Vis-Spectrometer)	PHERAstar FSX	BMG Labtech, Germany
Cytometer	CytoFlex S	Beckman-Coulter, Germany

2.1.2 List of chemicals

Table 2.2: Chemicals, enzymes and buffers utilized in this work.

Chemical	Supplier	Molecular formula
Isopropanol	Carl Roth, Germany	C_3H_8O
Dithiothreitol (DTT)	Carl Roth, Germany	$C_4H_{10}O_2S_2$

Table 2.2: Chemicals, enzymes and buffers utilized in this work (continued).

Chemical	Supplier	Molecular formula
Dimethylsulfoxide (DMSO)	Carl Roth, Germany	$C_2H_6OS \cdot 2H_2O$
Ethanol; ReagentPlus, 99.8%	Carl Roth, Germany	$C_2H_6O_2$
Glass-beads (0.25-0.5 mm)	Carl Roth, Germany	-
Glas beads (2.85-3.45 mm)	Carl Roth, Germany	-
Hydrochloric acid fuming	Carl Roth, Germany	HCl
n-Hexane	Carl Roth, Germany	C_6H_{14}
3,5-Dinitrosalycilic acid	Sigma-Aldrich, Germany	$C_7H_4N_2O_7$
Palmitic acid	Carl Roth, Germany	$C_{16}H_{32}O_2$
Glycerol trioleate	Sigma-Aldrich, Germany	$C_{57}H_{104}O_6$
Potassium sodium tartrate tetrahydrate	Sigma-Aldrich, Germany	$C_4H_{12}KNaO_{10}$
Sodium hydroxyde	Carl Roth, Germany	NaOH
Potassium dihydrogen phosphate	Carl Roth, Germany	KH_2PO_4
Sybr Green I	Invitrogen™, Thermo Fisher Scientific, Germany	-
Sybr Green II	Invitrogen™, Thermo Fisher Scientific, Germany	-
D(-)-Fructose	Carl Roth, Germany	$C_6H_{12}O_6$
D(+)-Saccharose	AppliChem, Germany	$C_{12}H_{22}O_{11}$
Tris-(hydroxymethyl)-aminomethan – TRIZMA	Sigma-Aldrich, Germany	$C_4H_{11}NO_3$
Tris hydrochloride	Carl Roth, Germany	$C_4H_{11}NO_3$ HCl
Potassium L-Glutamate	Sigma-Aldrich, Germany	$C_5H_8KNO_4$ H_2O

Cultivation

Chemical	Supplier	Molecular formula
Agar-Agar, Kobe I	Carl Roth, Germany	-
Nutrient Broth No. 4	Sigma-Aldrich, Germany	-
LB broth (10 g/L NaCl)	Sigma-Aldrich, Germany	-
LB broth (5 g/L NaCl)	Sigma-Aldrich, Germany	-
Yeast extract	Carl Roth, Germany	-
Caseinhydrolysate, standard	Carl Roth, Germany	-
Ammonium chloride	Carl Roth, Germany	NH_4Cl

Table 2.2: Chemicals, enzymes and buffers utilized in this work (continued).

Chemical	Supplier	Molecular formula
Ammonium iron(III) citrate	Honeywell	$C_6H_8O_7Fe_3NO_7$
Copper(II) sulfate pentahydrate	Carl Roth, Germany	$CuSO_4\ 5H_2O$
D(+)-Glucose	Sigma-Aldrich, Germany	$C_6H_{12}O_6$
Di-Potassium hydrogen phosphate	Carl Roth, Germany	K_2PO_4
Iron(III) chloride hexahydrate	Carl Roth, Germany	$FeCl_3\ 6H_2O$
Magnesium chloride	Carl Roth, Germany	$MgCl_2$
Magnesium sulphate monohydrate	Carl Roth, Germany	$MgSO_4\ H_2O$
Manganese(II) sulphate monohydrate	Carl Roth, Germany	$MnSO_4\ H_2O$
Manganese(II) chlorid tetrahydrate		$MnCl_2\ 4H_2O$
Iron(II) sulfate heptahydrate	Carl Roth, Germany	$FeSO_4\ 7H_2O$
Calcium nitrate tetrahydrate	Sigma-Aldrich, Germany	$Ca(NO_3)_2\ 4H_2O$
Thiamine hydrochloride	abcr GmbH, Germany	$C_{12}H_{17}ClN_{40}S$
Zinc sulphate heptahydrate	-	$ZnSO_4\ 7H_2O$
Potassium chloride	Carl Roth, Germany	KCl
Antibiotics		
Ampicillin sodium salt	Carl Roth, Germany	$C_{16}H_{18}N_3NaO_4S$
Kanamycin sulphate	Carl Roth, Germany	$C_{18}H_{36}N_4O_{11}\ 7H_2SO_4$
Spectinomycin dihydrochlorid pentahydrat	Sigma-Aldrich, Germany	$C_{14}H_{24}N_2O_7\ 2HCl\ 5H_2O$
Zeocin	InvivoGen, Germany	$C_55H_86N_2O0_21S_2$
Molecular biology		-
Agarose standard	Carl Roth, Germany	-
CutSmart buffer	New England Biolabs, USA	-
1 kbp DNA ladder	Carl Roth, Germany	-
DNA Stain Clear G	SERVA Electrophoresis GmbH Germany	-

Table 2.2: Chemicals, enzymes and buffers utilized in this work (continued).

Chemical	Supplier	Molecular formula
Reaction buffer C	New England Biolabs, USA	-
Enzymes		
Albumin Fraction V (BSA)	Carl Roth, Germany	-
Ampligase®	Lucigen, USA	-
DpnI	New England Biolabs, USA	-
Hybrid DNA Polymerase	Roboklon GmbH, Germany	-
OptiTaq DNA Polymerase	EURx, Poland	-
Q5®High-Fidelity DNA Polymerase	EURx, Poland	-
Q5®Reaction Buffer	EURx, Poland	-
T4 DNA Ligase	New England Biolabs, USA	-
T4-polynucleotidekinase	New England Biolabs, USA	-
Taq DNA Polymerase	New England Biolabs, USA	-
Lysozyme	Sigma-Aldrich, Germany	-
Lipase, immobilized on Immobead 150 from *Rhizopus oryzae* ($\geq$300 U/mg)	Sigma-Aldrich, Germany	-
Antibodies		
Monoclonal antibody	Anti-6His Tag monoclonal antibody, Alexa Fluor 488	Thermo Fisher Scientific, Germany
Analytical standards GC		
n-Dodecane	Carl Roth, Germany	$C_{12}H_{26}$
Heptadecane	Fluka, Germany	$C_{17}H_{36}$
Pentadecane	Fluka, Germany	$C_{15}H_{32}$
Glyceryl trioleate	Sigma-Aldrich, Germany	$C_{57}H_{104}O_6$
Oleic acid	Carl Roth, Germany	$C_{18}H_{34}O_2$

2.1.3 List of consumables

Table 2.3: List of consumables and kits employed in the present study.

Consumable	Type	Supplier
Microtiter plate	96 Well F-Bottom with condensation rings LUMITRAC^{TM} 200	Greiner Bio-One GmbH, Germany
Cell culture microplate	Cell culture 24 Well plate	Greiner Cellstar®
Cuvettes	Polystyrene cuvettes	Sarstedt
-	Cuvettes electroporation	Biozyme
Microcentrifuge tubes (1.5, 2 mL)	Eppendorf tubes	Sarstedt, Germany
HPLC column	Luna silica (5 μm, 100 x 4.6 mm)	Phenomenex, Germany
Centrifugal filters	Polyethersulfone membrane with 10 K MWCO	VWR, Germany
Centrifuge tubes	Polypropylene 15, 50 mL	Sarstedt, Germany
PCR tubes & Strips	-	Biorad, Germany
Screw cap microtubes	2 mL SC Micro tube	Sarstedt, Germany
Sterile filter medium	Filtropur S 0.2/0.45 μm	Sarstedt, Germany
Sterile syringes	-	Braun
Glass vials for GC/HPLC analysis	ND9, ND13	Th. Geyer, Germany
Micro-inserts for ND9 vials	-	Th. Geyer, Germany
Pipetting tips	Biosphere Filter tips (1000-100-10 μL)	Sarstedt, Germany
Kits		
PCR clean up	High pure PCR template	Roche Diagnostics GmbH, Germany
	Nucleospin Gel and PCR clean up	Macherey-Nagel GmbH & Co., Germany

Table 2.3: List of consumables and kits employed in the present study (continued).

Consumable	Type	Supplier
	Monarch® PCR & DNA Cleanup Kit	New England Bio-labs, USA
Genomic DNA purification	Monarch® DNA Gel Extraction Kit	New England Bio-labs, USA
Plasmid DNA pudrification	Monarch® Plasmid Miniprep Kit	New England Bio-labs, USA
	High Pure plasmid isolation kit	Roche Diagnostics GmbH, Germany
	innuPREP Plasmid Mini kit2.0	Analytik Jena, Germany
Rolling cyrcle amplification (RCA)	Illustra GenomiPhi V2 DNA amplification kit	GE Healthcare, UK
Sequencing	Mix2Seq Kit	Eurofins Genomics, Germany

2.1.4   Softwares

Table 2.4: Software for processing of GC/HPLC, data sets analysis, sequence design, text editor.

Software	Version & Packages, Manufacturer	Utility
CyBio Composer	v02.66.02.00 (Analytik Jena, Germany)	Robotic platform
Geneious	Geneious prime	Primer desgin & DNA sequence alignment
Inkscape	Version 1.0.1	Graphic editor
Labsolution	Version 5.92 (Shimadzu)	Chromatograph analyzer, peak assignment
Agilent ChemStation	Version - (Agilent)	Chromatograph analyzer, peak assignment
TeXstudio	v2.12.6	TeX distribution
Rstudio	1.1.453 (3.1.0 ggplot2; 0.8.2 tidyvers; 1.2.1 dplyr)	Statistics & data plotting
ZEN lite 2011	Carl Zeiss, Germany	Microscopy

2.2   Methods

2.2.1   Media

Media are routinely prepared as 1 L solutions in bottles by using desalted water (dH_2O). After preparation media are sterilized by autoclaving at 120 °C for 20 min.

2.2.1.1   Lysogeny broth (LB) media

LB^{10} medium was prepared by adding 25 g of LB (tryptone 10 g/L, yeast extract 5 g/L, NaCl 10 g/L) in 1 L of dH_2O, whereas LB^5 medium was prepared by adding 20 g of LB (tryptone 10 g/L, yeast extract 5 g/L, NaCl 5 g/L) in 1 L of dH_2O.

2.2.1.2   Difco sporulation medium (DSM)

Medium was prepared by mixing nutrient broth No.4 8 g/L, $MgSO_4{\cdot}7H_2O$ 0.25 g/L and KCl 1 g/L. Following medium was autoclaved. After cooling down the following sterile filtered components were added: 1 mL of 1 M $Ca(NO_3)_2$, 1 mL of 10 mM $MnCl_2$, 1 mL of 1mM $FeSO_4$

2.2.1.3   2 x Schaeffer's glucose (2 x SG) sporulation medium

Medium was prepared by mixing nutrient broth No.4 16 g/L, $MgSO_4 \cdot 7H_2O$ 0.5 g/L and KCl 2 g/L. Following medium was autoclaved. After cooling down the following sterile filtered components were added: 1 mL of 1 M $Ca(NO_3)$, 1 mL of 0.1 M $MnCl_2$, 1 mL of 1 mM $FeSO_4$, 2 mL of 50 % w/v glucose.

2.2.1.4   *Bacillus subtilis* transformation (TRAFO) medium

Transformation medium was prepared by mixing 10 mL of 10 x MN medium (0.137 g/mL dipotassium hydrogen phosphate, 60 mg/mL potassium dihydrogen phosphate, 8.4 mg/mL tri-sodium citrate) with 5 mL of 40 % D-glucose, 500 μL of 40 % potassium glutamate, 500 μL of 2.2 mg/mL ammonium iron citrate and 300 μL of 1 M magnesium sulfate. The final volume of medium was adjusted to 100 mL by adding 84 mL of sterile Type I H_2O.

2.2.1.5   *Bacillus subtilis* expression mix

2.5 % v/v yeast extract mixed with 2.5 % v/v casamino acids (CAA) in Type I H_2O.

2.2.1.6   Agar plates

Agar plates were prepared by adding 15 g/L of agar in the base medium.

2.2.2   Buffers

2.2.2.1   10 x Phosphate-buffered saline (10 x PBS)

80 g of NaCl, 2 g of KCl, 14.4 g of Na_2HPO_4, and 2.4 g of KH_2PO_4 were dissolved in 800 mL of dH_2O. Following, pH was adjusted to 7.4. Final volume was adjusted to 1 L with additional dH_2O.

2.2.2.2   Potassium phosphate (KPi) buffer, 60 mM, pH 6.4

Buffer was prepared by dissolving 5.737 g of KH_2PO_4 and 3.109 g of K_2HPO_4 in 950 mL of dH_2O. pH was adjusted to 6.4 with addition of KOH. Final volume was adjusted to 1 L with additional dH_2O.

2.2.2.3   Tris acetate EDTA (TAE) buffer (50 x)

TAE buffer was commonly prepared in 50 x stock solution by dissolving 242 g of Tris base in dH_2O adding 57.1 mL glacial acetic acid and 100 mL of 500 mM EDTA (pH 8). Final volume was adjusted to 1 L by addition of dH_2O.

2.2.2.4   Tris-HCl buffer, 100 mM, pH 7.0-9.0

Tris-HCl buffer was prepared by dissolving 6.057 g of Tris base in 450 mL of dH_2O. pH was adjusted to the desired value by addition of concentrated HCl 37% (12.1 M). Final volume was adjusted to 500 mL.

2.2.3   Microbial methods

2.2.3.1   Cultivation of *B. subtilis* in LB medium

Routinely, cultivation of *B. subtilis* was done in flasks with maximum volume of 50 mL or 200 mL in 37 °C with agitation 200 rpm.

2.2.3.2   Cryostock preparation

A single fresh colony was used to inoculate 4 mL LB medium containing the appropriate antibiotic. The culture was incubated at 37 °C with agitation 200 rpm for 12 h. After incubation, 800 μL of the culture were transferred to a cryostock tube containing 200 μL of 100 % glycerol. For every strain, two cryostock tubes were prepared. Tubes were stored in -80 °C and -20 °C freezer respectively.

2.2.3.3   Sporulation induction in *B. subtilis*

Sporulation was induced by the nutrient exhaustion method [86]. Fresh colonies of *B. subtilis* strains were used to inoculate 5-20 mL LB medium supplemented with 1 % w/v glucose containing the appropriate antibiotic. Cultures were incubated overnight (12-15 h) at 37 °C with agitation 200 rpm. After incubation cultures were harvested by centrifugation at 3,488 x g for 10 min. Cell pellet was resuspended in 2 x SG medium and inoculated into 500 mL Erlenmeyer flasks filled with 50-200 mL of prewarmed 2 x SG or medium containing the respective antibiotic. Cell cultures were incubated at 30 °C or 37 °C depending on the spores assay, for 48-56 h with shaking at 200 rpm. Sporulation progress was monitored daily, using a contrast phase microscope. When the amount of spores was sufficient, cultures were centrifuged at 3,488 x g for 15 min. Supernatant was removed whereas pellet was resuspended in 4 mL of sterile Type I water or the appropriate buffer depending on the assay. Spore suspensions were stored in the fridge.

2.2.3.4 *B. subtilis* spores purification using lysozyme

A spore culture of final volume 50 mL was centrifuged at 3,488 x g for 15 min. Supernatant was removed and pellet was resuspended in 10 mL of 75 μg/mL lysozyme in dH_2O, in order to lyse the cells. The suspension was incubated at 37 °C for 45 min with agitation at 200 rpm. After lysis, the suspension was centrifuged at 3,488 x g for 15 min at 4 °C. After centrifugation, supernatant was gently removed with a syringe. Pellet was resuspended in 10 mL of ice cold dH_2O or the appropriate buffer, and centrifuged as described above. The same process was repeated once more. After the last washing step, spore pellet was resuspended in 2 mL of ice cold dH_2O or the appropriate buffer.

2.2.4 Molecular biology methods

2.2.4.1 Agarose gel electrophorises

Agarose gels were prepared in gel casting chambers commonly using 1 % standard agarose and 1 x TAE Buffer. Gels were stained prior to casting using 2.5 μL SERVA DNA Stain Clear G per gel (50 mL) according to the manufacturer's recommendation. 1 kbp and 100 bp DNA ladder were used as reference size bands. Agarose gels were usually run for 35 min at 100 Volts. DNA bands were visualized using UV light and documented using a Gel Doc imager.

2.2.4.2 Amplification of DNA with Polymerase Chain Reaction (PCR)

All fragments used for the construction of plasmids were PCR amplified using the Hybrid Polymarase. Colony screening for correct assembly of the constructs was done with colony PCR (cPCR) by employing the Taq DNA polymerase. All PCRs were performed in a Tetrad 2 thermal cycler. Reaction mixture composition and thermocycler programs are listed in Tables 2.5 to 2.9.

Chapter 2. Common Materials & Methods

Table 2.5: PCR composition of DNA fragments amplified with Hybrid DNA polymerase.

Component	Volume/Reaction	Final concentration
10x Hybrid buffer containing 15 mM MgCl$_2$	5.0 μL	1x
Hybrid DNA polymerase 2 U/μL	0.5 μL	1 U
dNTP mix (10 mM)	1 μL	0.2 mM
Forward primer (10 μM)	5.0 μL	1 μM
Forward primer (10 μM)	5.0 μL	1 μM
DNA template	variable	<0.5 ng/50 μL
DMSO	0.5-2 μL	1-4 %
H$_2$O PCR grade	Variable	-
Total volume	50 μL	-

Table 2.6: cPCR composition of DNA fragments amplified with Taq DNA polymerase

Component	Volume/Reaction	Final concentration
10x C buffer	2.0 μL	1x
Taq DNA polymerase 5 U/μL	0.2 μL	1 U
dNTP mix (10 mM)	0.4 μL	0.2 mM
Forward primer (10 μM)	2.0 μL	1 μM
Forward primer (10 μM)	2.0 μL	1 μM
DNA template	5 μL of cell suspension	-
DMSO	0.2 μL	1 %
H$_2$O PCR grade	Variable	-
Total volume	20 μL	-

Table 2.7: Standard PCR thermocycling conditions for DNA fragments amplified with Hybrid DNA polymerase

Cycle step	Temperature [°C]	Duration [s]	
Initial denaturation	95	180	
Denaturation	95	30	
Annealing	55-65	30	× 35 cycles
Elongation	72	30s/kb	
Final elongation	72	2x elongation time	
Hold	15	-	

Table 2.8: Two step PCR thermocycling conditions for DNA fragments amplified using primers with overhangs and Hybrid DNA polymerase

Cycle step	Temperature [°C]	Duration [s]	
Initial denaturation	95	180	
Denaturation	95	30	
Annealing	55-58	30	× 5 cycles
Elongation	72	30s/kb	
Denaturation	95	30	
Annealing	63-65	30	× 30 cycles
Elongation	72	30s/kb	
Final elongation	72	2 x elongation time	
Hold	15	-	

Table 2.9: Standard cPCR thermocycling conditions for DNA fragments amplified with Taq DNA polymerase

Cycle step	Temperature [°C]	Duration [s]	
Initial denaturation	95	300	
Denaturation	95	30	
Annealing	55-65	30	× 30 cycles
Elongation	72	1min/kb	
Final elongation	72	2 x elongation time	
Hold	15	-	

2.2.4.3   Determination of DNA concentration

DNA concentration was determined using UV5nano spectrophotometer. Measurements were conducted by applying 1 μL of sample. For every sample absorbance at 230 nm (chaotropic salts), 260 nm (DNA) and 280 nm (proteins) was defined. According to these values the concentration of DNA was calculated automatically by the device. A typical value for protein purity ($A_{260/280}$) was between 1.7 and 2.0. Lower values indicated contaminations. A typical value for salt purity ($A_{260/230}$) was around 1.5.

2.2.4.4   Isolation of genomic DNA from *B. subtilis*

Genomic DNA of *B. subtilis* was extracted using the High Pure PCR Template kit according to the manufacturer's instructions.

2.2.4.5   Isolation of plasmid DNA from *(E. coli)*

Plasmid DNA from *E. coli* was extracted using the Monarch® Plasmid Miniprep kit or the innuPREP Plasmid Mini kit2.0, according to the manufacturer's instructions.

2.2.4.6   Transformation of *E. coli* with electroporation

An aliquote (50 μL) of *E. coli* electrocompetent cells (TOP10 or NEB 10) stored in -80 °C was thawed on ice prior to transformation. Cells were mixed with 10-15 μL of SLiCE mixture for ligation or 1 μL of plasmid with a concentration around 50 ng/μL for retransformation. In case of SLiCE ligation, 30 μL of ice cold Type I water were added to the transformation. The cell/DNA mix was transferred to a chilled electroporation cuvette without introducing bubbles. The cuvette was carefully whipped with a tissue and then was placed to the electroporator. Electroporation was conducted using the program Ec2 (25 kV) of the Bio-Rad MicroPulser™. Immediately after, 500 μL of LB medium were added to the cuvette and the whole mix was transferred to a 2 mL micro centrifuge tube. Cells were incubated at 37 °C with shaking at 200 rpm for 1 h. Following, different amounts of the transformation mix were plated on prewarmed LB plates containing the appropriate antibiotic. Plates were incubated overnight at 37 °C.

2.2.4.7 Transformation of *B. subtilis* with natural competence

Genomic integration and xylose induction for recycling of the antibiotic resistance cassette: For transformation, an overnight culture was prepared, containing 1mL of transformation medium (TM), 0.05% CAA and the appropriate antibiotic. The culture was inoculated with 15 cryostock of μL *B. subtilis* or preferably with a fresh colony. The culture was incubated overnight at 37 °C with shaking at 200 rpm. The next day 50 μL of the overnight culture were used to inoculate 950 μL of TM containing antibiotic. The fresh culture was incubated at 37 °C with shaking at 200 rpm for 6-8 h until the cells were competent. Competence was assessed by microscopy. Competent cells are much smaller in size and move really fast. At this point 1 μg of plasmid DNA was added to the culture and was further cultivated for 30 min. Afterwards, 200 μL of expression mix were added to the transformation for regeneration and cultivation continued for 1h. Finally different amounts of the transformation culture were plated on agar plates containing the antibiotic for selection. Plates were incubated at 37 °C incubator overnight to ensure formation of single colonies.

Recycling of the genomic integrated resistance cassette flanked by *lox66* and *lox71* regions, was achieved by induction of the Cre recombinase with xylose. A few (7-15) single colonies were resuspended in 300 μL of LB[5] containing 1% xylose and zeocin (20 μg/mL). Eppendorfs were incubated for 5-7 h at 37 °C with shaking at 200 rpm. In order to assure proper aeration, a hole was created on the lid of the 2 mL microcentrifuge tube. Induction of Cre recombinase results in recombination of *lox71* and *lox66* sites leading to excision of the resistance cassette placed in between and generating the scar *lox72*. After the induction, dilution streaking was performed on LB agar plates containing zeocin. Plates were incubated at 37 °C overnight. Single colonies were selected and excision of the resistance cassette was verified by colony PCR and by testing growth on negative control plates [131].

Transformation of low copy plasmid: For transformation of CopySwitch plasmid (p14035) the protocol was slightly modified. In that case, the plasmid was linearized by *XhoI* restriction enzyme and >200 fmol/mL were added to the transformation. Regeneration time was elongated to 4 h.

Transformation of high copy plasmid: For transformation of high copy plasmids (p02030, p02033), rolling circle amplification of the plasmids by the Phi-polymerase was required. Amplification was done according to the Illustra Genomi-Phi V2 DNA amplification kit. In brief, in a PCR tube 4.25 μL of sample buffer were mixed with 0.5 μL of plasmid (<5 ng). This mixture was incubated at 95 °C followed by a cooling step at 4 °C. Afterwards, 4.25 μL reaction buffer, 0.5 μL of Phi-polymerase and 0.5 μL of dNTPs were added in this mixture. The final reaction was incubated at 30 °C for 90 min. After the amplification, the whole mixture was added to the transformation culture and continued with the transformation protocol as described above. Regeneration time was 1 h.

2.2.4.8   SeamLess ligation cloning extract (SLiCE) cloning method

The SLiCE is a restriction-enzyme free *in vitro* cloning method, employed for assembly of several DNA fragments into a vector using *E. coli* extract. SLiCE facilitates fusion of the DNA fragments by recombining short end homologies [127]. In the present work SLiCE was used for construction of plasmids, assembling up to five DNA fragments. Each fragment was a PCR product containing 35 bp of homology to the side fragments. Fragments were mixed in a volume of 12 μL with a molar ration of vector to insert between 1:5 and 1:10 with an amount of vector around 0.1 pmol. 1.5 μL of 1x T4-DNA-Ligase reaction buffer and 1.5 μL of SLiCE extract (*E. coli* TOP10 cells), were added to the reaction. The mixture was incubated for 1 h at 37 °C . Afterwards, 10-15 μL of the mixture were added to 50 μL of *E. coli* electrocompetent cells. Transformation was conducted as described in Section 2.2.4.6.

2.2.4.9   Ligase cycling reaction (LCR)

The LCR is a scarless cloning method used for plasmid assembly, based on the hybridization of DNA fragments with complementary oligonucleotides called bridging oligos [95, 162]. The bridging oligos are used to connect neighboring DNA parts. In the current book LCR was employed for the assembly of up to four fragments. The bridging oligos were designed using the CloneFlow plugin available in Geneious. Each fragment was PCR amplified using phosphorylated oligos. Phosphorylation of the oligos was catalyzed by the T4-polynucleotidekinase. The reaction (50 μL) contained 4 μM of oligos, 4 mM adenosine triphosphate (ATP), 10 U of T4-polynucleotidekinase and 1x T4-polynucleotidekinase buffer. The reaction was incubated for 1 h at 37 °C followed by an inactivation step at 65 °C for 20 min. PCR amplification of the DNA fragments was conducted using 2.5 μL of the phosphorylated oligos in 50 μL of PCR. To increase the efficiency of LCR, the PCR amplified fragments were gel extracted in order to remove the excess amount of oligos.

For the assembly, the purified PCR fragments were mixed in a volume of 25 μL with a molar ratio of vector to insert 1:10. The final reaction was performed in a PCR tube and contained: 0.3 nM vector, 3 nM of each insert, 30 nM of each bridging oligo, 1x ampligase buffer (lelf made, without NAD^+, produced according to the composition of the ampligase buffer produced supplied by Lusigen, USA), 0.5 mM NAD^+, 10 mM $MgCl_2$, 0.45 M betaine, 8% v/v DMSO and 1.5 U Ampligase®. The mixture was incubated in a thermocycler (Tetrad 2 thermocycler; Bio-rad) using the cycling conditions shown in Table 2.10. Afterwards, 7.5 μL of the LCR mixture were used to transform 50 μL of electrocompetent *E. coli* cells as described in Section 2.2.4.6.

Table 2.10: LCR cycling conditions.

Cycle step	Temperature [°C]	Duration [s]
Initial denaturation	94	120
Denaturation	94	10
Annealing	55	30
Ligation	66	60
Hold	15	-

Denaturation, Annealing and Ligation steps: × 50 cycles.

3 Construction of germination mutant strains and development of tools for their analysis

3.1 Introduction

B. subtilis spores have great potential in industrial applications however their ability to resume vegetative life is one of the main reasons that hinder their widespread application. To improve their applicability, genome editing was employed and specific genes regarding germination were deleted via homologous recombination.

Bacterial spores can remain for long time metabolically inactive, given however the right trigger they are able to rapidly initiate macromolecular synthesis and metabolic activity in a process called germination followed by outgrowth [164]. Natural germinants are nutrients such as sugars and amino acids. The germinants bind on specific receptors located on the spore's inner membrane and initiate a cascade of reactions which involves first the excretion of Ca^{2+}-dipicolonic acid (CaDPA) stored in large amounts in the spore core and second the intake of water leading to its rehydration [167]. Germination completion is facilitated by the activation of cortex lytic enzymes (CLEs) responsible for the degradation of the spore cortex [164, 167].

In *B. subtillis*, two CLEs are implicated in the cortex hydrolysis and these are CwlJ and SleB [167]. These CLEs recognize muramic acid-δ-lactam (MAL) which can be found only in the cortex and not in the cell growing peptidoglycan (PG), and thus specifically degrade only the cortical PG. Spores lacking both lytic enzymes are unable to degrade their cortex and inevitably remain blocked in stage I of germination [79, 164]. Another important enzyme for the completion of germination is CwlD which encodes a N-acetylmuramoyl-L-alanine amidase involved in the synthesis of MAL [147]. In the absence of *cwlD* there is no MAL formation in the cortex consequently leading to inability of the CLEs to hydrolize the modified PG [147, 148].

Here in order to create germination deficient strains we deleted *cwlD* and *sleB* genes. To evaluate the effect of the knock out genes in the germination and outgrowth of the respective strains, different concentrations of purified spores were plated on LB plates. To define the germination rate it was essential to know the actual number of spores plated. Therefore a standard curve was made combining flow cytometry and reference beads, to correlate optical density to actual number of spores. Furthermore to enable quantitative and qualitative evaluation of different populations formed during sporulation, a flow cytometric method was developed employing DNA staining [86]. The presented method is a rapid technique especially suitable for quantification of germination deficient strains that can not be handled with conventional methods such as spore plating and colony counting. The developed method was tested in liquid sporulation cultures of germination deficient strains where the identity of the observed populations was validated by cell sorting and microscopy [86]. Finally in view of employing the developed method for high throughput analysis of sporulating strains under multiple conditions, a pilot experiment was designed using a robotic platform.

3.2   Materials & Methods

3.2.1   Strains & plasmids

All *B. subtilis* strains were developed as described in Section 2.2.4.7. Plasmids were assembled using the SLiCE cloning method Section 2.2.4.8 and transformed in *E. coli* according to Section 2.2.4.6. The oligonucleotides used for the plasmid construction as well as the plasmid maps are available in the supplemental.

All plasmids and oligonucleotides were designed using the Geneious Prime®software [88]. Sequencing was performed by utilizing the Mix2Seq kit supplied by Eurofins Genomics (Ebersberg, Germany).

Chapter 3. Construction of germination mutant strains and development of tools for their analysis

Table 3.1: List of strains employed in the current work.

Identifier	Relevant genotype	Resistance	Ref.
***B. subtilis* strains**			
B. subtilis KO7 (Bs02001)	PY79 + ΔnprE ΔaprE Δepr Δmpr ΔnprB Δvpr Δbpr	-	*Bacillus Genetic Stock Center ID 1A1133*
Bs02002	Bs02001 + Δ*sacA::(ZeoR, PxylA-cre, Pspac-comS, lacI)*	*zeo*	
Bs02003	Bs02002 + Δ*cwlD::lox72* Δ*sleB::lox72*	*zeo*	This work
Bs02004	Bs02002 + Δ*cwlD::lox72*	*zeo*	This work
Bs02005	Bs02002 + Δ*sigE/spoIIGA::lox72*	*zeo*	This work
Bs02018	Bs02003 + Δ*pksX::(PcotYZ, cotB:linker:sfGFP, lox72)* Δ*cotB::lox72*	*zeo*	This work
Bs02020	Bs02003 + Δ*cotA::PsspB, mrgA*	*zeo*	This work
Bs02022	Bs02003 + Δ*skf::lox72*	*zeo*	This work
Bs02025	Bs02022 + Δ*spo0E::lox72*	*zeo*	This work
***E. coli* strains**			
E. coli TOP10	*F- mcrA Δ(mrr-hsdRMS-mcrBC) Φ80lacZΔM15 ΔlacX74 recA1 araD139 Δ(ara leu) 7697 galU galK rpsL (StrR) endA1 nupG*	*str*	Invitrogen, Germany

Table 3.1: List of strains employed in the current work (continued).

Identifier	Relevant genotype	Resistance	Ref.
E. coli NEB 10-beta	*Δ(ara-leu) 7697 araD139 fhuA ΔlacX74 galK16 galE15 e14-φ80dlacZΔM15 recA1 relA1 endA1 nupG rpsL (StrR) rph spoT1 Δ(mrr-hsdRMS-mcrBC)*	*str*	NEB, USA

Table 3.2: List of plasmids employed in the current work.

Identifier	Relevant genotype	Resistance	Ref.
pJK196	*E. coli* plasmid used for the insertion of *ZeoR,PxylA-cre,Pspac-comS,lacI* in the *sacA* locus. Used for the expression of Cre recombinase. Development of strain Bs02002.	*amp*	PhD book of J. Kabisch
p02002 (pJET-cwlD)	*E. coli* plasmid developed for the deletion of *cwlD*. Contains two homology regions upstream and downstream of *cwlD*. Used for the development of strain Bs02004.	*spec*	This work
p02005 (pJET-sleB)	*E. coli* plasmid developed for the deletion of *sleB*. Contains two homology regions upstream and downstream of *sleB*. Used for the development of strain Bs02003.	*spec*	This work

Chapter 3. Construction of germination mutant strains and development of tools for their analysis

Table 3.2: List of plasmids employed in the current work (continued).

Identifier	Relevant genotype	Resistance	Ref.
p02004 (pJET-spoIIGA/sigE)	*E. coli* plasmid developed for the deletion of *spoIIGA/sigE*. Contains two homology regions upstream and downstream of *spoIIGA/sigE*. Used for the development of sporulation deficient strain Bs02005.	*spec*	This work
p02006 (pJET-cotB)	*E. coli* plasmid developed for the deletion of *cotB*. Contains two homology regions upstream and downstream of *cotB*. Used for the development of strain Bs02006.	*spec*	This work
p02130 (pKJ179-sfGFP)	*E. coli* plasmid developed for the integration of *PcotYZ:cotB:linker:sfGFP* fusion in the *pksX* locus. Contains two homology regions upstream and downstream of *pksX*. Used for the development of strain Bs02018.	*spec*	This work
p02019 (pJET-cotA)	*E. coli* plasmid developed for the deletion of *cotA*. Contains two homology regions upstream and downstream of *cotA*. *cotA* is exchanged by the *PsspB:mrgA* fusion gene. Used for the development of strain Bs02020.	*spec*	This work
p02023 (pJET-spo0E)	*E. coli* plasmid developed for the deletion of *spo0E*. Contains two homology regions upstream and downstream of *spo0E*. Used for the development of strain Bs02025.	*spec*	This work

Table 3.2: List of plasmids employed in the current work (continued).

Identifier	Relevant genotype	Resistance	Ref.
p11083.2	*E. coli* plasmid used for the deletion of *skfA*. Used for the development of strain Bs02022.	*spec*	PhD book of S. Hackenschmidt

3.2.2 Enumeration of spores using flow cytometry and counting beads

3.2.2.1 Sporulation and spore purification

For this experiment the germination mutant strain Bs02003 was employed. Spores were produced by the nutrient exhaustion method by cultivation in 2 x SG medium [137]. 50 μL of cryostock were used to inoculate 25 mL of 2 x SG medium containing the antibiotic zeocin (20 μg/mL) (Section 2.2.1). Cells were cultivated at 37 °C with agitation (200 rpm) for 48 h. Sporulation progress was monitored via microscopy. After 48 h a sufficient amount of spores was obtained in the cultures. The cultivation was stopped and the spores were purified by using the lysozyme treatment protocol (Section 2.2.3.4). PBS was used as washing buffer.

3.2.2.2 Enumeration of spores

Samples containing purified spores with $OD_{600\,nm}$ between 0.1 and 0.8 were prepared. Optical density of the samples was measured by BMG PheraSTAR FSX plate reader. 195 μL of the samples were transferred in a 96 well F-bottom plate. For washing purposes, between wells containing samples there was one well filled with PBS. In each of the sample wells, were added 50 μL of 8 peak beads with a defined concentration of 10^7/mL (Sony). Afterwards samples were analyzed using Beckman-Coulter CytoFlex S equipped with a 488 nm Argon Ion laser. FSC gain was set at 1,500 and FSC threshold was set at 75,000. Events were gated according to FSC-H/FSC-W. Measurement was conducted with a slow flow rate of 10 μL/min. Acquisition was stopped after counting 10,000 beads or measuring for 5 min. Spores concentration was calculated according to Equation (3.1).

$$Spores/mL = \frac{(\text{No. of spores}_{counted})(\text{No. of beads}_{added})}{(\text{No. of beads}_{counted})(\text{Volume}_{suspension}\text{ in mL})} \tag{3.1}$$

The resulting data were correlated to the measured $OD_{600\,nm}$ values and were used to create an enumeration/OD standard curve.

3.2.3   *B. subtilis* spores germination assay

Strains Bs02002, Bs02003 and Bs02004 were tested for their efficiency to germinate. Strains Bs02003 and Bs02004 harbor deletions of *sleB* and *cwlD* genes whereas Bs02002 is a germinating strain (Table 3.1). Cryostocks (50 μL) of each strain were used to inoculate 25 mL 2 x SG medium. Cells were cultivated at 37 °C with shaking at 200 rpm for 48 h. Following, cultures were centrifuged at 3,488 x g for 15 min. Residual cells were lysed and spores were purified using lysozyme according to the protocol in Section 2.2.3.4. Afterwards for each strain triplicates of spore suspensions in dH_2O with $OD_{600\,nm} \sim 0.1$ and final volume 0.4 mL were prepared. Optical density was measured in a 96 well polysterene F-bottom microplate (Greiner Bio-one GmbH, Germany) using the microplate reader BMG PheraSTAR FSX. The employed method included 1 min shaking at 600 rpm and measurement from the top of the plate. The spore suspensions were heat activated at 85 °C for 25 min and then a series of dilutions were prepared for each strain as shown in Table 3.3. 400 μL from each dilution were plated on LB[5] agar plates containing zeocin (20 μg/mL). Plates were incubated at 37 °C overnight. Germination rate (%) was calculated according to Equation (3.2). The No. of plated spores in stock was estimated using the spores enumeration standard curve (Section 3.2.2) which correlates optical density to spores/mL.

$$\text{Germination rate (\%)} = \frac{\text{CFU}}{\text{No. of plated spores}} 100 \qquad (3.2)$$

Table 3.3: Recommended dilutions for germination assay. The term "undiluted" refers to the initial stock with $OD_{600\,nm} \sim 0.1$ and final volume 0.4 mL.

Strain	No. of spores ($\times 10^6$) in stock	Dilutions after heat shock
Bs02002	(10.3 ± 1.1)	10^{-3}, 10^{-5}, 10^{-7}
Bs02003	(9.0 ± 0.4)	undiluted, 10^{-2}, 10^{-4}
Bs02004	(8.8 ± 0.9)	undiluted, 10^{-2}, 10^{-4}

3.2.4 Separation and quantification of spores using fluorescent staining and flow cytometry

3.2.4.1 Cell cultivation

Strain Bs02005 was cultivated in 4 mL LB[5] (Section 2.2.1) containing the antibiotic zeocin (20 μg/mL). Cultivation was done at 37 °C with shaking at 200 rpm for 14 h.

3.2.4.2 Spore production and purification

Sporulation was done by the nutrient exhaustion method [137]. The selected *B. subtilis* strains were inoculated in 4 mL LB[5] (Section 2.2.1) containing the antibiotic zeocin (20 μg/mL). Cultivation was done at 37 °C with shaking at 200 rpm for 14 h. Afterwards, cells were centrifuged at 3,400 x g for 10 min. Supernatant was removed while cell pellets were used to inoculate 200 mL of DSM (Section 2.2.1) containing the antibiotic zeocin (20 μg/mL). Sporulation was done in 500 mL Erlenmeyer flasks. Cultivation was done at 37 °C for 72 h. Afterwards the spore cultures were transferred in 50 mL centrifugal tubes and the spores and cells were harvested by centrifugation at 3,488 x g for 15 min. Following, purification of spores was conducted with Renografin gradient [137]. Renografin (sodium diatrizoate) is a reagent with high osmolality. Separation of spores from cells takes place due to different density and buoyancy [152].

The pellet from each centrifugal tube was resuspended in 2 mL sterile dH$_2$O. The suspension was transferred to a 2 mL opaque micro centrifuge tube and centrifuged at 16,000 x g for 10 min. Supernatant was once again removed and the pellet was resuspended in 2 mL of 20% w/v Renografin in Type I water. 10 mL of 50% w/v Renografin in Type I water were added in a 15 mL falcon covered with aluminum foil. The suspension was gently pippetted on top of that in order to form two distinct layers. The gradient was centrifuged at 11,000 x g for 30 min. After centrifugation, the supernatant was carefully removed and the formed pellet containing spores, was washed three times with 2 mL of cold sterile Type I water each (centrifugation at 11,000 x g for 10 min). After the final washing step, pellet was resuspended in 1 mL cold sterile Type I water. Purity was validated via contrast phase microscopy or flow cytometry.

3.2.4.3 Staining of spores and cells with fluorescent dyes

Two different fluorescent dyes were utilized for quantification of the spores: SYBR green I (SYBR1) (10,000 x stock, Invitrogen™) and SYBR Green II (SYBR2) (10,000 x stock, Invitrogen™). Generally SYBR green dyes bind to nucleic acids, whereas SYBR2 has higher sensitivity for RNA [65].

SYBR1 and SYBR2 stocks were diluted 1:100 times (100 x stock) with Type I water respectively. For each dye, aliquots of working solutions were transferred in 1.5 mL opaque tubes; aliquots of working dilutions were kept at 4 °C. The dyes were tested in different concentrations as well as in combinations. The final tested concentrations used for analysis of SYBR1 and SYBR2 ranged from 1 x to 4 x of the suppliers suggested concentration. Prior to flow cytometry, 500 μL aliquots of cells and spores at an $OD_{600\,nm}$ of 0.5 in 1 x PBS, were prepared. After re-suspension, the samples were mixed with 5 μL, 10 μL or 20 μL of each dye working solution respectively and were incubated in room temperature in the dark for 20 min [86].

3.2.4.4 Sample preparation for flow cytometric analysis of *B. subtilis* cultures

For each strain precultures were set from cryostocks by streaking the stock solution onto LB agar plates containing zeocin (20 μg/mL). Plates were incubated overnight at 37 °C. The next day a single colony was picked and inoculated into a tube containing 4 mL of LB medium with zeocin (20 μg/mL). Cell cultures were grown overnight at 37°C with agitation (200 rpm). Afterwards cells were harvested by centrifugation at 16,000 x g for 1 min. The supernatant was removed and the cell pellets were resuspended in 2 mL DSM. The cell density of the resuspensions was defined photometrically at 600 nm and an amount of each suspension was used to inoculate 30 mL prewarmed of DSM containing the antibiotic zeocin (20 μg/mL). The initial $OD_{600\,nm}$ of the cultures was 0.1. Cultivation was done at 37 °C with shaking at 200 rpm. After 48 h of cultivation in the sporulation medium, samples of 200 μL from each culture were acquired for flow cytometry. Samples were centrifuged at 16,000 x g for 1 min with subsequent removal of the supernatant. Pelleted cells were resuspended in 500 μL 1 x PBS and stained with 2 x SYBR1 as mentioned above. The same process was repeated at 48, 56, 72, 80 and 96 h. All experiments were conducted in three biological replicates.

3.2.4.5 Flow cytometry and FACS

Cytometric analyses were conducted using a Sony LE-SH800SZBCPL with a 488 nm argon laser. Gains for photomultipliers for the channels SSC and FL-1 (525/50 nm) were set on 40.0 %, 39.0 % and 51.0 % respectively with a FSC-threshold of 0.20 % and a window extension of 50. The FSC diode was set on an amplification level of 16/16 and sample pressure was set so that events per second (eps) were kept under 30,000. For each analysis, 10^5 events were evaluated.

For cell sorting a 100 μm microfluidics sorting chip for Sony SH800 was used. The same settings as for cytometric analyses were employed, however eps were kept under 12,000.

3.2.4.6 Microscopic analysis of sorted populations

Microscopic pictures were captured using an inverted microscope. For these experiments, 5 x 10^6 events from the selected populations were sorted in 15 mL tubes coated with 5 % BSA. Sorted samples were transferred in 20 mL centrifugal filters of 30 kDa molecular cutoff and centrifuged at 3,400 x g for 5 min. Agarose pads were prepared on glass slides with 1 % agarose standard. Microscopy slides were prepared by pipetting 3 μL of the concentrated sample suspensions on top of the agarose pads placed on glass slides and were coveried with glass coverslips. Samples were observed with Axio Vert.A1 (Carl Zeiss, Germany) inverted microscope equipped with 100 x oil immersion objective and 10 x ocular magnification. Images were captured with AxioCam ICm1 (Carl Zeiss, Germany) camera using ZEN lite 2011 software (Carl Zeiss, Germany).

3.2.4.7 Viability assessment after staining

To monitor the effect of staining in the viability of the cells and spores, distinct samples of cells from the sporulation deficient strain Bs02005 and spores from the germinating strain Bs02002 were stained with SYBR1 and SYBR2 as mentioned above. Viability was assessed by sorting 210 events from each sample on an LB plate containing zeocin (20 μg/mL). Sorting was performed with the single cell three drops mode. Cells and spores were sorted on the plate according to the 384 well plate sort layout. Subsequently, plates were incubated overnight at 37 °C. For analysis, the relative number of colonies in respect to the total number of sorted events was evaluated.

3.2.5 High-throughput system for quantification of *B. subtilis* spores

To develop a high throughput system for quantification of spores, a custom made robotic platform configured by Analytic Jena, was employed. The platform was equipped with a four axis laboratory robot Precis Flex 760 (Precise Automation, US-CA) working on a linear rail, a labware storage carousel including one rack for 22 microtiter plates, a pipetting-robot CyBio FeliX 96 Channel, the microplate reader BMG PheraSTAR FSX, an automated incubator Cytomat2 (Thermo Fisher Scientific, US-MA) and a flow cytometer Beckman-Coulter CytoFlex S.

Process design and programming of the robotic platform was done by the PhD student Thomas Zoll.

3.2.5.1 Cultivation of *B. subtilis* cells in 24 well plates containing LB medium

Pre-cultures were prepared using LB medium. *B. subtilis* strains were cultivated in 750 μL LB[5] containing zeocin (20 μg/mL) in Cellstar 24-well plates (Greiner Bio-one GmbH, Germany). Each well was inoculated with a fresh colony. For each strain, quadruplicates were prepared and inoculation was done following a randomized plate design to exclude positioning effects. The plate was placed on top of a wet towel in a container. The wet towel was used to reduce medium evaporation. The container was incubated overnight at 37 °C with agitation at 200 rpm. After cultivation, $OD_{600\ nm}$ was measured using a plate reader.

3.2.5.2 Sporulation of *B. subtilis* cells in 24 well plates containing 2 x SG medium

Sporulation of the strains was conducted in Cellstar 24-well plates with 750 μL 2 x SG medium containing the appropriate antibiotic. The cultures were inoculated with cells cultivated overnight in LB as mentioned above. The initial OD $_{600\ nm}$ was 0.1. The plate was placed in the incubator of the robotic platform and incubated for 48 h at 37 °C with agitation at 150 rpm. $OD_{600\ nm}$ was measured hourly starting at 0 h.

3.2.5.3 Preparation of dilution and staining plates

A LUMITRAC™ 200, 96 well plate (Greiner Bio-one GmbH, Germany) was used for storage of the SYBR1 fluorescent dye. One of the wells was filled with 100

μL of stain of 1x concentration prepared as mentioned in section 2.2.8. The lid of the plate was covered with brown tape to stimulate dark conditions. For every flow cytometric measurement, a LUMITRAC™ 200, 96 well plate was employed containing 190 μL of PBS. These plates were used for dilution and staining. The final volume of the dilution was 200 μL. The lids of the plates were covered with brown tape as well.

3.2.5.4   Quantification of distinct populations in sporulating cultures using flow cytometer

Cytometric analysis was conducted using a Beckman-Coulter Cytoflex S equipped with a 488 nm argon laser. Gains for the channels SSC, FSC and FITC were set on 1'500, 1'100 and 76 respectively. Flow rate was set at 10 μL/min. Acquisition was terminated after analyzing 150 μL of sample or measuring for 120 sec. Events were gated according to FSC-H/FITC. Data analysis was done using Rstudio.

In Table 3.4 are presented the steps executed for high throughput quantification of sporulation efficiency.

Table 3.4: Workflow of high-throughput quantification of spores in the robotic platform.

Step	Action	Device	Repetition
1	$OD_{600\,nm}$ measurement: Plate with sporulation cultures in 2 x SG medium (Plate 1)	BMG PheraSTAR FSX	hourly (t=0...48)
2	Dilution: 10 μL from each cultivation well (Plate 1) were transferred in the dilution plate containing 90 μL PBS (Plate 2)	CyBio FeliX 96 channels	every 6 hours (t=0...48)
3	Staining: 2 μL of SYBR1 were pipetted in each one of the wells in Plate 2. Plate was incubated for 20 min covered with the lid	CyBio FeliX 96 channels	every 6 hours (t=0...48)
4	Quantification: Plate 2 was measured in the cytometer	Beckman-Coulter CytoFlex S	every 6 hours (t=0...48)

3.3 Results & Discussion

3.3.1 Development of standard curve for enumeration of *B. subtilis* spores

To develop a spores enumeration standard curve, 16 samples of purified spores from strain Bs02003 were prepared with $OD_{600\,nm}$ ranging between 0.08 and 0.4. Higher optical densities were avoided to eliminate the risk of clogging the flow cytometer. Each sample contained spores purified with lysozyme and a defined number of reference beads as an internal standard. To exclude doublets and small multiplets or debris, data were analysed and gated according to forward scatter height (FSC-H) and forward scatter width (FSC-W) Results are shown in Figure 3.1.

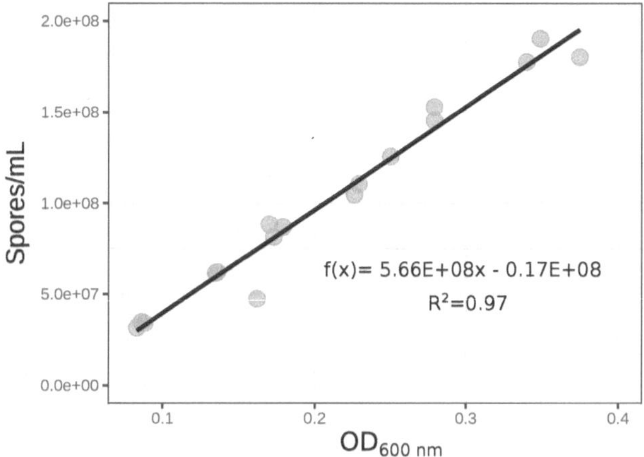

Figure 3.1: Enumeration of spores using flow cytometry and reference beads. Standard curve of spore counts (Spores/mL) correlated to optical density ($OD_{600\ nm}$).

Data generated from flow cytometry were correlated to optical density measurements to generate a standard curve. Fitting of the data was done following the linear regression model. The calculated equation and the estimated R-squared are shown in Figure 3.1. Purification of spores utilizing lysozyme even though is a very efficient method, has the drawback of cell debris accumulation in the sample. Cell debris scatter light and thus affect both optical density measurements and cytometric analysis. Hence this could lead to wrong calculation of actual spore counts. Using gradient centrifugation as a spore purification method instead, where cells are separated from spores due to different density, could bypass the aforementioned problems.

3.3.2 Development and analysis of *B. subtilis* germination deficient strains

For the purpose of generating germination deficient spores, two mutant strains were constructed. Strain Bs02004 harbors deletion of *cwlD* whereas strain Bs02003 is *cwlD sleB* double mutant. Both genes encode for germination specific cortex lytic enzymes [163, 109, 164]. Different concentrations ($OD_{600\ nm}$) of purified spores from each strain were heat activated for 25 min at 85°C and plated on LB plates. Optical density of plated spores was correlated to actual number of spores employing a spores enumeration standard curve shown in Figure 3.1. The germination efficiency was assessed by counting colony forming units (CFU) in reference to the number of spores plated. The germinating strain Bs02002 was used as a positive control. For each strain three technical replicates were prepared. Results are presented in Table 3.5.

Table 3.5: *B. subtilis* strains and the respective germination rates. All values are given with standard deviations.

Strain	Relevant genotype	Germination rate (%)
Bs02002	wild type	31 ± 7
Bs02003	$\Delta cwlD\ \Delta sleB$	$(1.56 \pm 0.14) \times 10^{-3}$
Bs02004	$\Delta cwlD$	$(1.9 \pm 0.14) \times 10^{-3}$

Colony forming units obtained for each strain revealed that both gene deletions have a key role in the completion of germination process. The estimated germination rate for both mutant strains was lower than 0.002 % whereas the germination rate for the control strain Bs02002 was $\sim$ 31 %. The obtained results are coincide with data published in previous studies [147, 79]. According to Setlow et al., what seems to prevent these mutants from completing germination, is the really low water content of the core which restrains stored enzymes from proper folding and function [164].

3.3.3 Development of flow cytometric method for separation of *B. subtilis* spores

3.3.3.1 Flow cytometric comparison of distinct cell and spore samples

To design a method for distinction of cells and spores, initially separate samples of vegetative cells from the non sporulating strain Bs02005 and purified spores from the germination mutant strain Bs02003 were employed. Samples were stained with two different DNA stains, SYBR1 and SYBR2 as described in Section 3.2.4.3. Following, they were analyzed by flow cytometry and compared with regard to their light scattering properties (SSC, FSC) and the fluorescence intensity (FI).

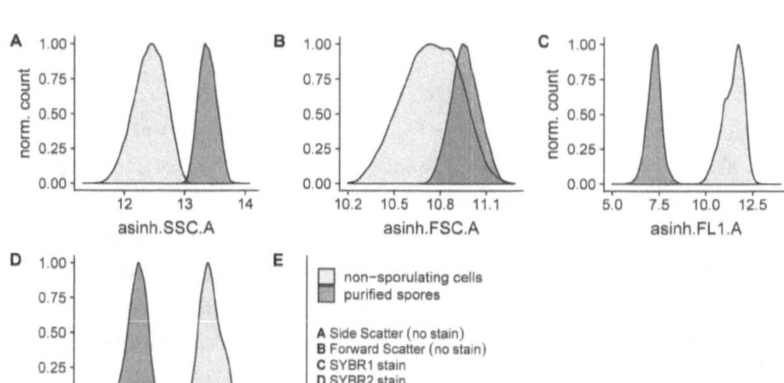

Figure 3.2: Density plots of FI and scatter signals of *B. subtilis* vegetative cells from strain Bs02005 and purified spores from strain Bs02003 measured separately [86].

As presented in Figure 3.2, spores exhibit higher SSC and FSC than vegetative cells (Figure 3.2A Figure 3.2B). Spores have a complex structure consisting of a thick peptidoglycan cortex encased by a multilayered proteinaceous coat [40]. It is assumed that the higher SSC and FSC obtained for the spores owes to their structural complexity. However the noted difference, is not enough to allow for clear differentiation of the two populations, especially when considering that during sporulation intermediate populations are formed such as cells containing forespores, which makes resolution more difficult. Hence, scattering channels alone are commonly not sufficient for distinction of all subpopulations and therefore fluorescence staining is employed as means for better separation. In contrast, Laflamme et al. showed that isolation of spores from *Bacillus subtilis var niger* was possible by employing UV induced autofluorescence with a less common UV laser [103]. Although the described method bypasses the need for an additional staining step, its employment is hampered by the use of special instrumentation.

Staining of the samples with SYBR1 and SYBR2 yields better resolution of the two populations (Figure 3.2C and Figure 3.2D). Vegetative cells have better membrane permeability compared to the spores, thus exhibit higher fluorescence intensity. In earlier studies, fluorescent staining has been successfully used to elucidate the sporulation dynamics of *Paenibacillus polymyxa* and *Clostridia* [29, 192].

3.3.3.2 Optimization of staining method by employing mixed samples of cells and spores

To assess the capacity of separating spores from cells co-existing in the same sample, a 1:1 artificial mix of vegetative cells and purified spores was stained with different concentrations of dye. Additionally different staining times were tested. Following this, cell viability was assessed by sorting a defined number of cells and spores on LB agar plates. In the current work Gaussian mixture modeling (GMM) was employed to separate the sub-populations of cells and spores. Routinely GMM is being used for prediction of sub-populations, but is commonly restricted to cases where signal distributions are at least near normal [13, 52]. As scattering and fluorescence signals in flow cytometry often follow log-normal distributions, thresholding or clustering with this method can be a useful tool for separation. The calculated distances between the distributions of the two populations, as predicted by GMM is shown in Figure 3.3 together with the pooled standard deviation.

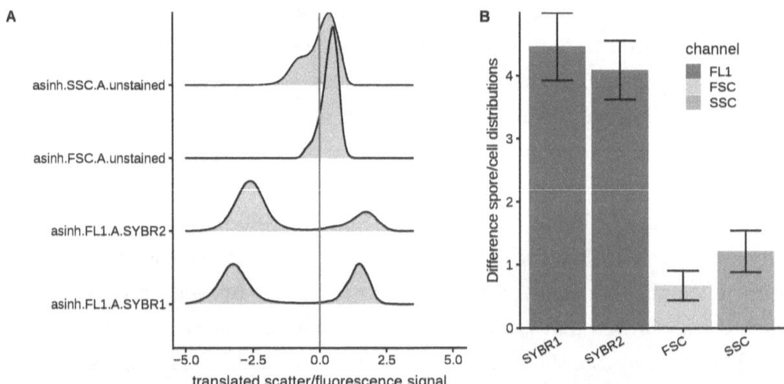

Figure 3.3: Optimization of staining method using artificial samples of 1:1 mixed cells and spores. Comparison of two different fluorescent dyes and channels utilized for separation. Staining conditions: 2 fold concentration, 20 minutes. **A.** Raw data and cutoff values as estimated by GMM. **B.** Calculated distance between distributions as evaluated by GMM in the respective channels. Error bars show the pooled standard deviation of three replicates. The same evaluation method was applied for the effect of staining time and stain concentration (Figure S1) [86].

Staining with SYBR1 resulted in the best resolution between the two populations, whereas minimum difference was observed when FSC was used for separation, which is in line with previous experiments. Variation of staining time as well as stain concentration yielded only minor differences (Figure S1A). Effects of the staining procedure on cell and spore viability was assessed by colony forming units as shown in Figure S1B. Since a 2-fold concentration of SYBR1 dye and 20 minutes incubation time seem to be sufficient for separation without significantly impairing cell viability, these conditions were employed for further investigations.

3.3.3.3 Identification and cell sorting of sub-populations obtained in sporulating cultures

The developed method was employed in order to assess whether it allowed for discrimination of intermediate populations formed during sporulation. Therefore a sporulating culture of strain Bs02003 was employed. After 48 hours of sporulation initiation, samples collected from the cultures, were stained with SYBR1 and were analyzed by flow cytometry. As shown in Figure 3.4 in the fluorescence channel two distinct populations were observed (Figure 3.4A). One of the two populations follows a bimodal distribution indicating the existence of two different sub-populations. The three populations show little differentiation in FSC and SSC (Figure 3.4B) The centers of the three clusters are shown in red. To identify the three observed populations FACS was employed and 5×10^6 events from each cluster were sorted. The identity of each cluster was validated by contrast phase microscopy. The three populations corresponded to cells, mother cells containing forespores and spores (Figure 3.4C).

Chapter 3. Construction of germination mutant strains and development of tools for their analysis

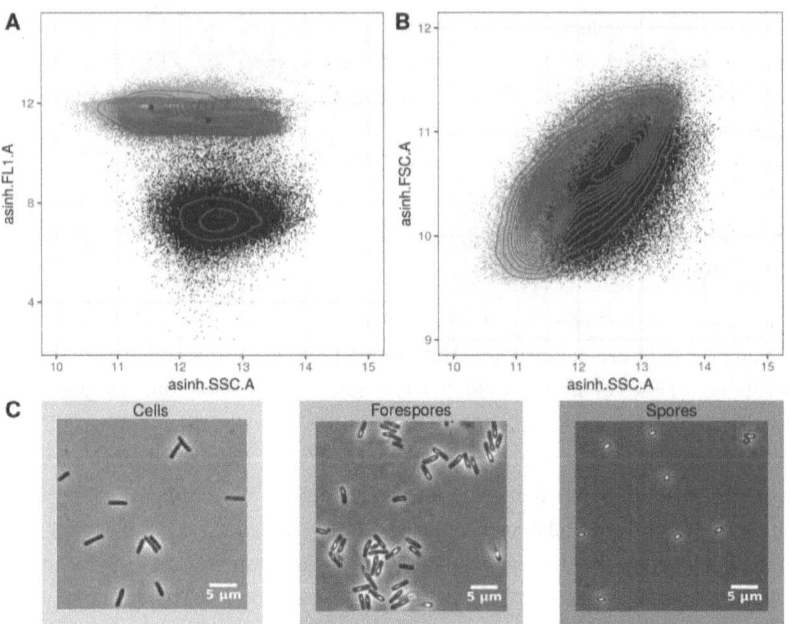

Figure 3.4: Classification of subpopulations in sporulating cultures. Hyperbolic-sine transformed side scatter (SSC) and fluorescence in FL1-A channel are shown in (**A**), whereas (**B**) shows front- and side scatter (FSC, SSC). The respective clusters as predicted by GMM, are indicated with different color. Cluster centers are shown in blue and red lines show 2d kernel densities. (**C**) Subpopulations were subsequently isolated by FACS. Corresponding microscopy images of the classified subpopulations are shown [86].

By employing GMM on a reference sample from Bs02003 strain containing cells, spores and forespores, the respective subpopulation of any other sample in the respective experimental set was classified by assigning each point to its closest center. In our case there was almost no shift of populations over time or among replicates. Consequently, Euclidean distance to the calculated centers was employed to classify subpopulations [86]. The three detected clusters were assigned to different stages of sporulation based on the following hypothesis: Vegetative cells have high permeability for SYBR1 and therefore exhibit high signal in the FL1 channel. By the time endospore formation is initiated, this permeability and FI is maintained in similar levels, however the emerging structure of the spore coat increases the granularity and hence the signal in the SSC channel. Furthermore, a small increase in FSC could be affiliated to an increase in the average cell size caused by intracellular spore coat formation. The transition from endospore formation to spore release, leads in cell size reduction and subsequently in lower FSC. As the spore formation is completed, permeability for SYBR1 is reduced in the presence of the multilayered proteinaceous coat, leading to a drop in FI [118]. Microscopic identification of the respective subpopulations supports this hypothesis.

3.3.3.4 Monitoring sporulation dynamics

The developed method was employed for quantitative assessment of sporulation dynamics of *B. subtilis* strains harboring different genomic modifications. All employed strains harbor deletions of *cwlD* and *sleB* leading to a germination deficient phenotype [163, 109, 164]. Strain Bs02018 is utilized for spore display of sfGFP as a fusion to the spore coat protein CotY whereas strain Bs02020 further harbors deletion of *cotA* which is responsible for the brownish pigmentation of spores [77]. Strain Bs2025 contains deletions of the genes *skfA* and *spoOe*. SpoOE is a phosphatase which dephosphorylates the master regulator of sporulation SpoOA~P and delays the process of sporulation whereas spore killing factor SkfA, is part of the *skf* operon responsible for the production of toxin during sporulation [144, 62]. Initially the sporulation progress of strain Bs02018 used for spore display of sfGFP, was analysed and compared to two other strains (Bs02003, Bs02020).

Sporulation progress of all strains at distinct time points, was monitored by determining population sizes as described above. In all cases, three distinct populations were observed as presented in Figure 3.5. A clear difference in the sporulation pattern between strain displaying sfGFP (Bs02018) and the two other non-display strains is visible, with Bs02018 exhibiting lower sporulation efficiency.

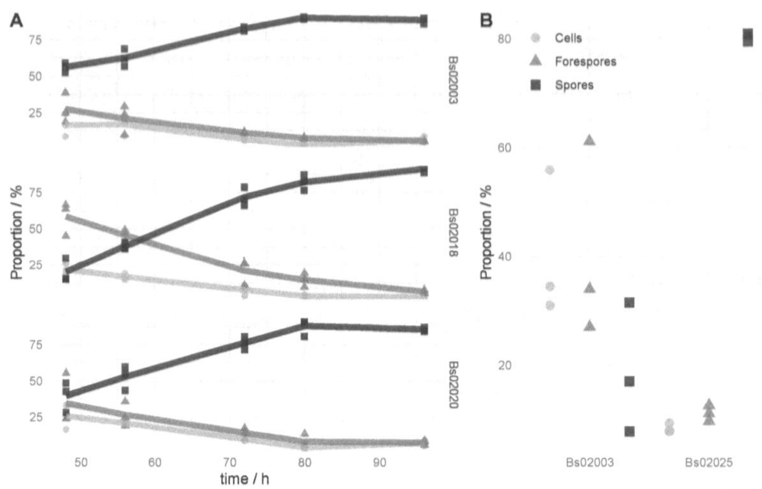

Figure 3.5: Sporulation dynamics of *B. subtilis* strains with different genomic modifications. Shift of culture composition in cells, spores and cells containing forespores over time is shown. **A.** Sporulation progress over the course of 96 hours for strains Bs02003, Bs02018 and Bs02020. All three strains are double mutants for *cwlD* and *sleB* whereas strain Bs02018 additionally expresses sfGFP as spore coat fusion protein. Strain Bs02020 additionally harbors deletion of *cotA*. **B.** Comparison of sporulation dynamics of two *B. subtilis* strains after 24 hours of cultivation. Strain Bs02025 harbors deletions of *cwlD*, *sleB*, *skfA* and *spoOE*, whereas Bs02003 is used as control. The selected strains and methods were further used for the development of a high throughput system employing a robotic platform. Results of the subpopulations composition over time are shown in Figure S4 [86].

As shown in Figure 3.5A after 48 hours of sporulation, 59 % of events in cultures of strain Bs02003 are spores in contrast to 35 % and 44 % for strains Bs02018 and Bs02020. In the course of 96 hours, maximum percentages of spores obtained for the three strains were 88 %, 91 % and 85 % respectively. It is observed that for all strains, only a minority of events represented cells without endospores. Furthermore for strain Bs02018 utilized for display of sfGFP a reduction in sporulation efficiency is obtained compared to the two other strains. It is assumed that the expression of sfGFP posed additional metabolic burden to the already energy consuming process of sporulation thus impairing sporulation efficiency.

The method was further used to evaluate the effect of *spoOE* and *skfA* deletions in strain Bs02025. Sporulation efficiency of Bs02025 was compared to the control strain Bs02003 (Figure 3.5B). After 24 hours of sporulation, cultures of Bs02025 already consisted of 80 % spores while only 18 % of events were classified as spores for the control strain. These findings are in line with results reported for *spo0E* and suggest that the selected mutations increased the sporulation efficiency [145].

In an effort to create a high throughput system which would allow analysis of sporulation dynamics for multiple strains under different cultivation conditions, a custom made robotic platform configured by Analytik Jena was employed. As a pilot, strains Bs02025 and Bs02003 were cultivated in 24 well plates containing 2 x SG media and analyzed with the here presented method. Strains were analyzed in regard to growth and sporulation efficiency. Results shown in Section 6.3 are in agreement with the findings obtained in the current experimental set.

3.4   Conclusion and outlook

The remarkable properties of spores have secured their utility in many biotechnological applications. Although germination is the key process to their revival, when it comes to spores utilization in industrial processes it is considered a fundamental issue. Germination of spores in a very high rate, occurs in response to high levels of certain amino acids. However, Sturm and Dworkin demonstrated that spores still can germinate, though with a reduced frequency, in the absence of any known signal [184]. Working towards the expansion of spore applications, the construction of germination deficient strains and development of new tools for rapid assessment of sporulation efficiency remains pivotal. By means of molecular biology and genome editing spore properties can be optimized to reach the desired results.

Here it was shown that double knockout of two genes necessary for germination, *cwlD* and *sleB*, considerably decreased the germination rate of strain Bs02003 to less than 0.001%. Germination efficiency was assessed by combining flow cytometry and counting beads with plate counting. To address the lack of available tools for quantification of sporulation dynamics in germination deficient strains, herein a flow cytometric method was developed employing fluorescent staining which allow for rapid quantification and isolation of cell and spore populations. The presented method was utilized for monitoring sporulation over time in germination mutant strains with different genomic modifications. For each culture three distinct populations were identified: cells, cells containing forespores and spores. A decrease in the sporulation efficiency of strain Bs02018, utilized for the display of sfGFP on the spores surface was observed. On the contrary, strain

Bs02025 which is double knock-out mutant of the phosphatase gene encoding Spo0E and of the spore killing factor SkfA exhibited the highest sporulation efficiency, as within 24 hours of cultivation in sporulation medium, cultures already consisted of 80 % spores as opposed to 18 % for the control strain.

Although reduction of germination rate was to a certain extend successful, still the question remains whether is possible to utterly hinder spores from germinating and outgrowing. Germination is a revival mechanism whereby a cell proliferates and repopulates a habitat. Thus, strategies which counteract this process most likely are not favored from natural selection. Recently Cutting and coworkers reported the construction of recombinant spores deprived of the ability to outgrow, albeit still capable of germinating [72]. The authors suggested that by introducing a thymine auxotrophy to the vegetative cells of *B. subtilis* they were able to prevent germinating spores from outgrowing in the absence of thymine. Although the results regarding outgrowth of spores were satisfying, the applicability of the method is hampered by the poor cell growth, consequently followed by reduction in the overall spores yield. One way to alleviate this problem could be the design of an inducible system where the "switch" to auxotrophy would be controlled by sporulation. This way the vegetative cells would grow normally prior to sporulation.

4 Development of *B. subtilis* spore surface display platform towards biocatalysis

4.1 Introduction

Enzymes are complex catalysts that differ in their structure and properties. Over the years, different immobilization methods have been developed in order to enhance the operational stability of the individual catalysts and reduce the cost of biocatalytic reactions by recycling of the enzymes [169]. Ideally the formulated biocatalyst should have a low cost of production and not leach. To this end employment of spore display fulfills most of the above mentioned criteria, since immobilization on the spore coat is based on fusion protein technology and is carried out concomitantly with protein expression. In order to evaluate the capacity of spores as biocarriers, a set of different enzymes were selected for immobilization via the fusion protein technology. First, different anchoring proteins and spacers were evaluated as putative fusion partners using fluorescent proteins. The selected anchoring proteins differ in structure and location on the spore coat. Following, the performance of spores in biocatalytic reactions was assessed by displaying different enzymes: a light inducible decarboxylase and two variants of sucrose phosphorylase. The selected enzymes differ in size, cofactor requirement and number of subunits.

4.1.1 *Cv*FAP: a light inducible fatty acid decarboxylase

In 2017 Beisson and co-workers reported on a flavin dependent photoenzyme named *Cv*FAP (PDB:5NCC) deriving from the microalga *Chlorella variabilis* which catalyzes the photocatalytic decarboxylation of free fatty acids to the corresponding n-1 alkanes and alkenes activated by blue light (450 nm) [181]. Crystal structure of *Cv*FAP revealed that the enzyme forms a monomer (63 kDa) harboring flavin adenine dinucleotide (FAD) cofactor as the light excited part of the protein. Since then various attempts have been made to use *Cv*FAP for the production of hydrocarbons employing different expression organisms [75, 204, 117, 15]. In the majority of the studies *Cv*FAP was expressed in *E. coli* and was used either in the form of purified enzyme or as cell free extract. Bruder et al. reported for the first time the expression of *Cv*FAP in the oleaginous yeast *Yarrowia lipolytica* [15]. Encouraged by these findings, we expressed *Cv*FAP in *B. subtilis* using the spore display approach in order to unravel the potentials of spores in the production of biofuels.

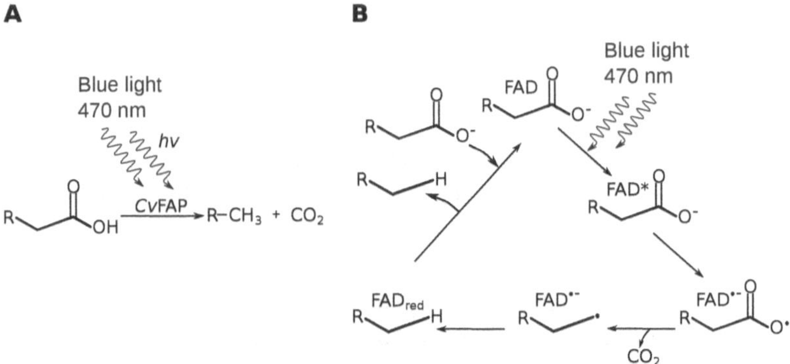

Figure 4.1: Light driven decarboxylation of fatty acids catalysed by *Cv*FAP. **A:** Simplified reaction scheme. **B:** Proposed mechanism of FAD photocycle. Scheme adapted from [104].

One part of the current work was focused on the investigation of the optimal reaction conditions for the production of hydrocarbons from the displayed *Cv*FAP. According to previous publications, *Cv*FAP exhibits a catalytic preference for long chain saturated fatty acids therefore palmitic acid was used as substrate [181, 75]. After determining the optimum conditions for the activity and stability of the displayed enzyme, the next aim was the application of spores in the production of drop-in biofuels from natural triglycerides (TAGs). Therefore an enzymatic cascade for the production of hydrocarbons from microbial oils was reconstituted. The biotransformation consists of a two step enzymatic cascade. In the first step a lipase is employed for the hydrolysis of lipids with subsequent release of free fatty acids (FFAs) and the second step includes the irreversible conversion of FFAs to the corresponding n-1 alkanes and alkenes catalysed by the spore displayed light inducible decarboxylase.

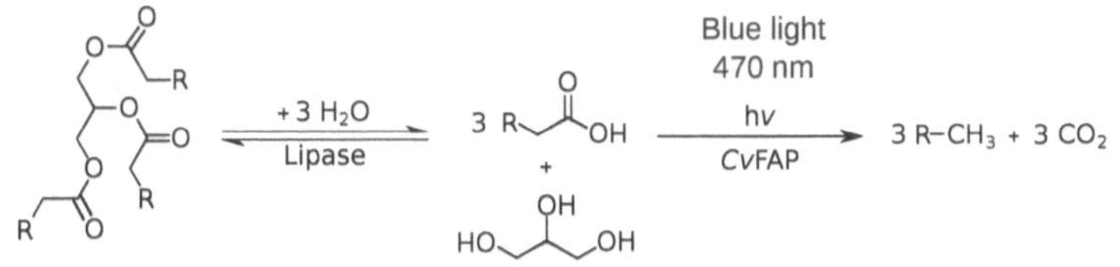

Figure 4.2: Two step enzymatic cascade for the production of hydrocarbons from triglycerides. Scheme adapted from [117].

4.1.2 Sucrose phosphorylase

Sucrose phosphorylases (SucP) are transglucosidases (EC 2.4.1.7) that belong to glycoside hydrolases (GH) family 13. SucP catalyzes the reversible phosphorolysis of sucrose to D-fructose and α-D-glucose-1-phosphate (G1P) by following a double displacement mechanism [116]. This reaction involves the attack of a carboxylic residue of the enzyme to the anomeric carbon of sucrose, resulting in the formation of a covalent β-glycosyl-enzyme intermediate. In the next step this intermediate can be displaced by phosphate. A downside of this process is that water can also displace the glycosylated intermediate leading to the hydrolysis of the substrate [116]. SucP can also catalyze the transfer of glycosyl moiety from G1P or sucrose to a variate of substrates such as monosaccharides, sugar alcohols, phenolic compounds and carboxylic acids (Figure 4.3) [33, 2, 14]. SucP mainly is found in lactic acid bacteria and bifidobacteria where it holds an important role in the energy metabolism via the formation of G1P [19]. Furthermore, G1P is an efficient donor for chemical and enzymatic glycosylation reactions therefore can be used in a variety of applications such as substitute of inorganic phosphate in parental nutrition [60, 120]. The commercial exploitation of SucP mainly relies on the low cost and abundance of sucrose used as substrate for the production of fine chemicals [61].

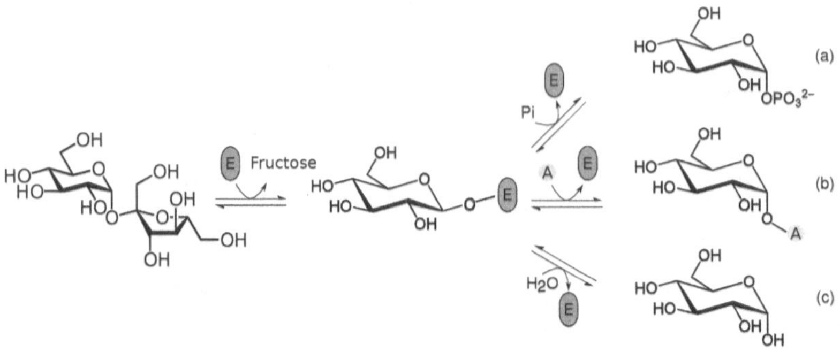

Figure 4.3: Reactions catalyzed by sucrose phosphorylase using sucrose as substrate: (**a**) Phosphorolysis, (**b**) Tranglycosylasion, (**c**) Hydrolysis (E: enzyme, A: acceptor, Pi: inorganic phosphate). Scheme adapted from [53].

Most of the industrial processes involving carbohydrate conversion are performed at 60 °C or higher, in order to avoid microbial contamination. So far sucrose phosphorylases have been isolated from mesophilic bacteria such as *Leuconostoc mesenteroides, Bifidobacterium adolescentis* and *Lactobacillus acidophilus* thus have limited thermostability at elevated temperatures. Hence, efforts are being made to increase the enzymatic stability at industrial conditions by means of enzyme immobilization or protein engineering [19, 20].

In the present book two different variants of sucrocrose phosphorylase were employed for spore display and tested for their efficiency to convert sucrose to G1P and fructose. The one variant derives from *Leuconostoc mesenteroides* (*Lm*SucP) and forms a monomer with a size of 56 kDa while the other variant derives from *Bifidobacterium adolescentis* (*Ba*SucP) and forms a dimer where each sub-unit has a size of approximately 56 kDa [94, 182].

4.2 Materials & Methods

4.2.1 Strains & plasmids

All *B. subtilis* strains were developed as described in Section 2.2.4.7. Plasmid p02022 was assembled using LCR cloning method (Section 2.2.4.9). All other plasmids were assembled using the SLiCE cloning method (Section 2.2.4.8) and transformed in *E. coli* according to Section 2.2.4.6. All information regarding PCR, oligos and plasmid maps are available in the supplemental.

All plasmids and oligonucleotides were designed using the Geneious Prime® software [88]. Sequencing was performed by utilizing the Mix2Seq kit supplied by Eurofins Genomics (Ebersberg, Germany).

Table 4.1: List of strains employed in the current work.

Identifier	Relevant genotype	Resistance	Ref.
B. subtilis strains			
B. subtilis KO7 (Bs02001)	PY79 + ΔnprE ΔaprE Δepr Δmpr ΔnprB Δvpr Δbpr	-	*Bacillus Genetic Stock Center ID 1A1133*
Bs02002	Bs02001 + Δ*sacA::(ZeoR, PxylA-cre, Pspac-comS, lacI)*	*zeo*	This work
Bs02003	Bs02002 + Δ*cwlD::lox72 ΔsleB::lox72*	*zeo*	This work
Bs02024	Bs02003 + Δ*pksX::(PcotYZ, cotZ:linker: LmSucP, lox72)*	*zeo*	This work
Bs02028	Bs02024 + Δ*cotZ::lox72*	*zeo*	This work
Bs02029	Bs02003 + Δ*cotY::lox72*	*zeo*	This work
Bs02031	Bs02006 + Δ*pksX::(PcotYZ, cotB:linker: LmSucP, lox72)*	*zeo*	This work
Bs02034	Bs17003 + p14035	*kan*	This work
Bs02035	Bs02003 + Δ*pksX::(PcotYZ, cotY:linker: CvFAP, lox72)*	*zeo*	This work

Table 4.1: List of strains employed in the current work (continued).

Identifier	Relevant genotype	Resistance	Ref.
Bs02037	Bs02035 + p14035	*kan*	This work
Bs02039	Bs02035 + p02030	*kan*	This work
Bs02040	Bs02029 + p02033	*kan*	This work
Bs17001	Bs02007 + *ΔpksX::(PcotYZ, cotG:flex linker:gfpmut2, cotB:flex linker:mkate2, lox72)*	*zeo*	Bsc book of D. Klewinghaus
Bs17002	Bs02007 + *ΔpksX::(PcotYZ, cotG:rigid linker:gfpmut2, cotB:rigid linker:mkate2, lox72)*	*zeo*	Bsc book of D. Klewinghaus
Bs17003	Bs02003 + *ΔpksX::(PcotYZ, cotY:flex linker:gfpmut2, cotZ:flex linker:mkate2, lox72)*	*zeo*	Bsc book of D. Klewinghaus
Bs17004	Bs02003 + *ΔpksX::(PcotYZ, cotY:rigid linker:gfpmut2, cotZ:rigid linker:mkate2, lox72)*	*zeo*	Bsc book of D. Klewinghaus
Bs17005	Bs02003 + *ΔpksX::(PcotYZ, cotG:rigid linker:gfpmut2, cotZ:flex linker:mkate2, lox72)*	*zeo*	Bsc book of D. Klewinghaus
Bs95002	Bs02003 + p95002	*kan*	Bsc book of P. Gockel

Table 4.1: List of strains employed in the current work (continued).

Identifier	Relevant genotype	Resistance	Ref.
E. coli strains			
E. coli TOP10	*F- mcrA Δ(mrr-hsdRMS-mcrBC) Φ80lacZΔM15 ΔlacX74 recA1 araD139 Δ(ara leu) 7697 galU galK rpsL (StrR) endA1 nupG*	*str*	Invitrogen, Germany
E. coli NEB 10-beta	*Δ(ara-leu) 7697 araD139 fhuA ΔlacX74 galK16 galE15 e14- φ80dlacZΔM15 recA1 relA1 endA1 nupG rpsL (StrR) rph spoT1 Δ(mrr-hsdRMS-mcrBC)*	*str*	NEB, USA

Table 4.2: List of plasmids employed in the current work.

Identifier	Description	Resistance	Source/Ref.
p02010 (pUC57-CvFAP)	Template plasmid for amplification of CvFAP codon optimized for *B. subtilis*	*amp*	BaseClear B.V
p02012 (pJET-cotZ)	*E. coli* plasmid developed for the deletion of *cotZ*. Contains two homology regions upstream and downstream of *cotZ*. Used for the development of strain Bs02028.	*spec*	This work
p02017 (pUC57-LmSucP)	Template plasmid for amplification of *Lm*SucP codon optimized for *B. subtilis*	*amp*	BaseClear B.V

Chapter 4. Development of B. subtilis spore surface display platform towards biocatalysis

Table 4.2: List of plasmids employed in the current work (continued).

Identifier	Description	Resistance	Source/Ref.
p02022 (pJK179-cotz:LmSucP)	*E. coli* plasmid developed for the insertion of *PcotYZ:cotZ:LmSucP* fusion gene in the *pksX* locus. Contains two homology regions upstream and downstream of *pksX*. Used for the development of strain Bs02024.	*spec*	This work
p02024 (pJET-cotY)	*E. coli* plasmid developed for the deletion of *cotY*. Contains two homology regions upstream and downstream of *cotY*.Used for the development of strain Bs02029.	*spec*	This work
p02025 (pJK179-cotB:LmSucP)	*E. coli* plasmid developed for the insertion of *PcotYZ:cotB:LmSucP* fusion gene in the *pksX* locus. Contains two homology regions upstream and downstream of *pksX*. Used for the development of strain Bs02031.	*spec*	This work
p02027 (pJK179-cotY:CvFAP)	*E. coli* plasmid developed for the insertion of *PcotYZ:cotY:CvFAP* fusion gene in the *pksX* locus.Contains two homology regions upstream and downstream of *pksX* locus. Used for the development of strain Bs02035.	*spec*	This work
p14035 (plcBReT-sh)	*E. coli* shuttle plasmid, contains **low copy** *ori* for *B. subtilis* and two homology regions upstream and downstream of *pksX* locus (2 kb each). Used for the development of strains Bs02034/37.	*kan*	[131]

Table 4.2: List of plasmids employed in the current work (continued).

Identifier	Description	Resistance	Source/Ref.
p14037 (pMSE3)	*E. coli* shuttle plasmid, contains **high copy** *ori* for *B. subtilis*. Used as template for PCR amplification of the backbone (pMSE3).	*kan*	PhD book of Florian Nadler
p02030 (pMSE3-cotY::CvFAP)	*E. coli* shuttle plasmid, contains **high copy** *ori* for *B. subtilis* and the fusion gene for *PcotYZ,cotY::CvFAP*. Used for the development of strain Bs02039.	*kan*	This work
p02033 (pMSE3-cotY::BaSucP)	*E. coli* shuttle plasmid, contains **high copy** *ori* for *B. subtilis* and the fusion gene for *PcotYZ,cotY::BaSucP*. Used for the development of strain Bs02040.	*kan*	Twist Bioscience
p95002 (pMSE3-cotY::RO lipase)	*E. coli* shuttle plasmid, contains **high copy** *ori* for *B. subtilis* and the fusion gene for *PcotYZ,cotY::RO lipase(R. oryzae)*. Used for the development of strain Bs95002.	*kan*	BSc book of P. Gockel

4.2.2 Fluorescence intensity (FI) measurement of spores displaying GFPmut2 and mKate2

4.2.2.1 Spores production and purification

Pre-cultures were prepared using LB medium. *B. subtilis* strains were cultivated in 600 μL LB[10] in Cellstar 24-well plates. Each well was inoculated with a fresh colony. For each strain, quadraplicates were prepared and inoculation was done following a randomized plate design to exclude positioning effects. The plate was placed on top of a wet towel in a container. The wet towel was used to reduce medium evaporation. The container was incubated overnight at 37 °C with agitation at 200 rpm.

Sporulation of the strains was conducted in Cellstar 24-well plates filled with 550 μL 2 x SG medium. The cultures were inoculated with 50 μL of overnight LB[10] culture as mentioned above following the same randomized plate pattern. Cells were cultivated for 48 hours at 37 °C with agitation at 200 rpm.

Spores were purified using the lysozyme treatment protocol. Finally spores were resuspended in 1 x PBS buffer.

4.2.2.2 Plate reader measurements of fluorescent intensity (FI) and optical density (OD) of purified spores displaying different fluorescent proteins

In order to conduct the measurements different spore dilutions in 1 x PBS were prepared. For the measurements a 96 well plate F-bottom was employed, filled with 50 μL of spore suspension. Measurements were conducted using the plate reader CLARIOstar (BMG Labtech). Settings for GFP FI were 470 nm excitation and 515 nm emission whereas for mKate2 excitation was set at 588 nm and emission at 633 nm. Measurements were performed in top reading mode. Optical density measurements were performed at 450 nm. Prior to all measurements a shaking step was included at 300 rpm for 30 sec. Data analysis and statistical tests were performed with R studio.

4.2.3 Preparation of photocatalytic reactions for spores displaying *Cv*FAP

4.2.3.1 Spores production and purification

Sporulation was carried out as described in Section 2.2.3.3. Cells where cultivated in 2 x SG (400 mL) at 30 °C for 48 h. Afterwards spores were purified using the lysozyme treatment protocol (Section 2.2.3.4). The lysis step was performed at 30 °C. Washing steps were done with Tris buffer (50 mM, pH 8). After the final washing step the whole spore pellet was resuspended in 2 mL Tris buffer (100 mM, pH 8.5). A spore dilution (1:100, 1 mL) was prepared in order to measure the optical density. The purified spores were used directly for the enzymatic reactions.

4.2.3.2 Spore mediated decarboxylation of palmitic acid in small scale

For the small scale photocatalytic reactions, 5 mL transparent glass vials were utilized (ND13, Sarstedt). Purified spores of $OD_{600\,nm} \sim 40$ (equal to 2.8 mg of spores dry weight) were resuspended in 300 μL Tris-HCl buffer (100 mM, pH 8.5) and were mixed with 300 μL palmitic acid dissolved in n-hexane containing 1 mM of dodecane, unless noted otherwise. The mixture was vortexed for 30 sec in order ensure formation of emulsions. Following, the vials were placed in a rotor covered with a drum with blue LEDs (Fig. 2.1). Vials were stirred at 70 rpm and illuminated continuously with blue light (130 mAmp, 4 V) for 20 h unless noted otherwise. Optimum incubation temperature was 30 °C. Reactions were terminated by switching off the light supply. Afterwards, the reactions were stored in -20 °C until extraction of the alkanes.

Figure 4.4: Custom made photocatalytic set up for small scale production of hydrocarbons.

4.2.3.3 Determination of kinetic parameters for the spore displayed *Cv*FAP

In order to determine the kinetic parameters (K_m, V_{max}) of the displayed enzyme, different concentrations (0.5-10 mM) of palmitic acid in n-hexane containing 1 mM of dodecane as internal standard were prepared. The final reaction mix contained 100 μL of substrate in n-hexane and 300 μL of catalyst in 100 mM Tris-HCl buffer pH 8.5. The catalyst was employed in a final concentration of round 5 mg/mL of SDW which corresponded to $OD_{600\,nm} \sim 40$. Substrate and catalyst were freshly prepared. The reaction vials were vortexed for 5 sec and then placed in the rotor. Reactions started by activating the blue light source (130 mAmp, 4 V, 68 μmol quanta m^{-2} s^{-1}). Vials were incubated at 30 °C and constantly stirred at 70 rpm. The reaction rate for each substrate concentration was recorded in distinct time points for 4 hours. For each time point distinct reactions were prepared. All reactions were performed in quadruplicates. The reactions were terminated by removing the samples from the light source and placing them in -20 °C. Extraction of alkanes followed and GC analysis was performed. Data outcome was analyzed by using R studio following the Michaelis-Menten model. Kinetic constants were calculated according to Equation (4.1) and Equation (4.2).

$$V_0 = \frac{V_{max}[S]}{[S] + K_M} \tag{4.1}$$

$$SDWsp.activity = \frac{V_{max}}{m_{SDW}} \tag{4.2}$$

4.2.3.4 Biotransformation of TAGs to HCs in one pot reaction

A stock of triolein dissolved in n-hexane was prepared to a final concentration of 0.2 M.

One-step biotransformation (1 mL): The final reaction consisted of 10 μL triolein from the stock solution or 25 μL oil (*Yarrowia lipolytica, C. oleaginosus* , olive oil), 50 mg of immobilized lipase from *Rhizopus oryzae* (Sigma-Aldrich) and 990 or 975 μL of spore suspension in 100 mM Tris-HCl, pH 8.5. The spore suspension in the reaction had an $OD_{600\ nm}$ of 120. Reactions were placed in the rotor and were incubated at 30 °C under constant stirring at 70 rpm for 24 hours.

Two-step biotransformation (1 mL): The first step of the reaction consisted of 10 μL triolein from the stock solution, 50 mg of immobilized lipase from *Rhizopus oryzae* (RO lipase) or alternatively spores from strain Bs95002 (Bsc book of Peter Gockel) displaying the lipase from *Rhizopus oryzae* with an OD_{600} of 160. The reaction volume in the first step was 700 μL. The hydrolysis reaction took place at 30 °C with agitation at 70 rpm for 24 hours when RO lipase was employed or 48 hours in the case of spore displayed lipase. Afterwards, 300 μL of Bs02039 spore suspension with a final $OD_{600\ nm}$ of 120 were added in the mixture and the reaction was further incubated for 24 hours in the presence of blue light.

In both types of biotransformation, the reaction vials were mildly vortexed prior to incubation. Reactions where *Cv*FAP was present, started by activating the blue light source (130 mAmp, 4 V, 68 μmol quanta m^{-2} s^{-1}). Reactions were terminated by removing the samples from the light source and placing them in -20 °C. Extraction of alkanes followed and GC analysis was performed. As negative control, strain Bs02003 was employed. To elucidate the possibility of TAGs hydrolysis in the presence of spores, reactions without the addition of lipase were performed. All reactions were performed at least in triplicates. Data were analyzed by using R studio.

4.2.3.5 Extraction of alkanes

For the extraction of the alkanes, reactions were transferred to 2 mL extraction tubes containing 0.35 g of glass beads (0.25-0.5 mm). For the reactions with palmitic acid as substrate, the mixtures were supplemented with n-hexane so the minimum volume of the organic phase was 300 μL. For the reactions with triolein and oils, extraction was done using 300 μL n-hexane containing 1 mM dodecane. Next samples vortexed vigorously using a ball mill (Mixer Mill MM 400) at 30 Hz for 15 min. Afterwards the mixture was centrifuged at 16,000 x g for 10 min and the upper organic phase was transferred to a glass vial for GC analysis.

4.2.3.6 Preparation of spore dry weight (SDW)

For SDW, a 150 mL sporulation culture was set. Spores were purified with lysozyme and were resuspended in 4 mL of the appropriate buffer. $OD_{600\,nm}$ was determined for the stock solution of the spores. 1 mL of the stock was transferred to a 2 mL micro centrifuge tube of determined weight. The suspension was centrifuged at 16,000xg for 10 min. The supernatant was removed whereas spore pellet was placed with open lid in a desiccator filled with silica beads for 48 h. Afterwards, SDW was determined gravimetrically.

4.2.3.7 Hydrocarbon analysis with gas chromatography (GC)

Detection of hydrocarbons produced in the reaction of spore presented *Cv*FAP with fatty acids, was performed using GC. The samples were analyzed with a Shimadzu Nexis GC 2030, on a Shimadzu SH-Rxi-5MS column (30 m, 0.25 mm, 0.25 μm) and detected by FID. The temperature of inlet and FID were set to 250 °C and 310 °C, respectively. The linear velocity of hydrogen was set to 50 cm/s. Split was set to 1. Column oven temperature program: initial temperature of 50 °C, which was held for 2.5 min, followed by a ramp to 250 °C at a rate of 10°C per min, followed by a ramp to 300 °C at a rate of 20 °C per min and a final step at 300 °C for 10 min.

Identification of hydrocarbon peaks was conducted using GC standards. For quantification of the corresponding peaks, a standard curve for each one of the hydrocarbons was prepared. Concentrations of the employed standards ranged between 0.025 mM and 1 mM. All standards were dissolved in n-hexane containing 1 mM dodecane as internal standard.

4.2.4 HPLC analysis of oils for identification of triglycerides and free fatty acids

Oils (Yali. oil, Co. oil, olive oi) were analysed with HPLC (Agilent 1260 infinity bio-inert, Agilent) equipped with a luna silica column (250 mm x 4.6 mm i.d., 5 μm particle size, Phenomenex) and a refractive index detector (Shodex-ri 101). The employed mobile phase was n-hexane:isporopanol (18:1). Flow was set to 1 ml/min and the analysis time was 20 min. Samples were prepared by mixing 1 μL of oil with 499 μL of mobile phase. Identification of peaks was conducted using analytical standards (triolein and oleci acid) in concentrations of 5 mM and 10 mM respectively. Standards were dissolved in mobile phase. Data analysis was conducted with R.

4.2.5 Immunofluorescence labeling of spores

Localization of spore presented proteins fused with a 6 x His tag was conducted via immunofluorescence. A suspension of purified spores treated with lysozyme, of $OD_{600} = 1$ was prepared. The suspension was centrifuged and spore pellet was washed three times with 1 mL PBS (pH 7.4) containing 1 % w/v BSA. After the last washing step, spore pellet was resuspended once more in 1 mL PBS (pH 7.4) containing 1 % w/v BSA and was incubated at room temperature for 30 min. After incubation, the suspension was centrifuged at 10,000 x g for 1 min. Afterwards supernatant was removed and pellet was resuspended in 500 μL of PBS (pH 7.4) containing 1 % w/v BSA. Resuspension was transferred to an opaque micro centrifuge tube and a volume of 5 μL of anti-6 x His-AlexaFluo[488] supplied from ThermoFisher (dilution 1:100) was pipetted in the mixture. The suspension was incubated on ice for 1h in the dark. After incubation, the suspension was washed three times with 500 μL of PBS (pH 7.4) containing 1% w/v BSA, each. All washing steps were performed with centrifugation at 8,000 x g for 5 min at 4 °C. After the last washing step, pellet was resuspended once again in 500 μL of the same buffer. For comparison reasons, experiments included wild type spores as well, which served as negative control. Spores were examined by fluorescence microscopy using Axio Vert.A1 (Carl Zeiss) inverted microscope equipped with

100 x oil immersion objective and 10 x ocular magnification. Green fluorescence was visualized using a standard filter (Ex = 490 nm, Em = 520 nm). Images were captured with AxioCam ICm1 (Carl Zeiss) camera using ZEN lite 2011 software (Carl Zeiss). Analysis of the images was done with ImageJ.

4.2.6   Production of D-fructose and G1P from spores displaying sucrose phosphorylase

4.2.6.1   Spores production and purification

Sporulation was carried out as described in Section 2.2.3.3. For the experiments employing the sucrose phosporylase from *Leuconostoc mesenteroides* (*Lm*SucP), strains Bs02028 and Bs02031 were cultivated in 2 x SG (400 mL) at 30 °C for 48 h. As control for native activity of sucrose phospohrylase, strain Bs02003 was employed. For the experiments employing the sucrose phosporylase from *Bifidobacterium adolescentis* (*Ba*SucP) strains Bs02043 and Bs02029 (control) were cultivated in 2 x SG (400 mL) at 37 °C for 48 h. Afterwards spores were purified using the lysozyme treatment protocol (Section 2.2.3.4). The lysis step was performed at the same temperature as the cultivation. For strains Bs02028, Bs02031 and Bs02003 (control), all washing steps as well as the final pellet resuspension were carried out with potassium phosphate buffer (60 mM, pH 6.4). For strain Bs02043 and Bs02029 all washing steps and the final pellet resuspension were carried out with potassium phosphate buffer (0.2 M, pH 7). After the final resuspension, a spore dilution (1:100, 1 mL) was prepared in order to measure the optical density.

4.2.6.2   Production of G1P from *Lm*SucP

For the reactions, 2 mL micro centrifuge tubes were utilized. Purified spores of $OD_{600\,nm}$ ranging between 10 and 40 were resuspended in 500 μL potassium phosphate buffer (60 mM, pH 6.4) containing 100 mM of D-sucrose. Reactions were incubated for 24 h at 30 °C with shaking at 200 rpm. Reactions were terminated by removing the spores from the reactions with centrifugation at 16,000xg for 5 min at 4 °C. Proteins and other residues were removed from the supernatant of the reaction by using centrifugal filters with 10 K molecular cutoff. Samples were stored at -20 °C until analysis.

4.2.6.3 Production of G1P from *BaSucP*

Reactions were performed in 2 mL micro centrifuge tubes. Reaction mixtures contained 0.2 M D-sucrose, 0.2 M potassium phosphate buffer pH 7 and spores in a final $OD_{600\,nm} \sim 0.7$. The final reaction volume was 2 mL. Reactions were initiated by addition of the spores. Mixtures were incubated for 4 h at 50 °C in a thermoshaker with shaking at 700 rpm. In distinct time points (0-5-10-15-30-60-120-150-240 min) samples of 100 μL were retrieved from the reaction mixture and were transferred in PCR tubes. Reactions were terminated by heat deactivation of the spore displayed enzyme at 80 °C for 5 min. Samples were stored on ice until DNS-assay. Reactions were performed in triplicates. One unit (1 U) was defined as the amount of spores that release 1 μmol fructose in 1 min at 50 °C and pH 7. The concentration of fructose corresponds to equimolar amount of G1P according to the reaction stoichiometry.

Storage stability was assessed by storing the spore displayed enzyme and the control spores at 4 °C in 0.2 M phosphate buffer, pH 7 for 102 days. Activity was measured using the DNS assay. Determination of the enzymatic activity was carried out by calculating the initial rate of the reaction as it was defined by the linear portion of the progress curve.

4.2.6.4 Reuse of displayed *BaSucP*

Reactions were performed in 2 mL micro centrifuge tubes. Reaction mixtures contained 0.1 M D-sucrose, 0.2 M potassium phosphate buffer pH 7 and spores in a final $OD_{600\,nm} \sim 7$. The final reaction volume was 1 mL. Reactions were initiated by addition of the spores. Mixtures were incubated for 15 min at 50 °C with shaking at 700 rpm. Afterwards, reactions were centrifuged at 11,000 x g for 1 min at 4 °C. 100 μL were retrieved from the supernatant and were transferred in PCR tubes for heat deactivation (80 °C for 5 min). Spore pellets were resuspended in 1 mL phosphate buffer 0.2 M, pH 7 and centrifuged once more. Supernatant was removed and spore pellets were resuspended in 1 mL of reaction mixture (0.1 M D-sucrose, 0.2 M potassium phosphate buffer pH 7). The same procedure was repeated for a total of 7 cycles. Reactions were performed in triplicates. Quantification of released fructose was conducted with the DNS assay as described above.

4.2.6.5 Quantification of reducing sugars with dinitrosalicylic acid (DNS) method

The DNS method is a colorimetric technique that consists of a redox reaction between the 3,5-dinitrosalicyclic acid and the reducing sugars present in the sample. The carbonyl group of these sugars can be oxidized to the carboxyl group by mild oxidizing agents, while the DNS (yellow) is reduced to 3-amino-5-nitrosalicylic acid (red-brown) [128].

1. **Preparation of DNS solution:** 50 mL of DNS solution were prepared by dissolving 0.375 g 3,5-dinitrosalicylic acid and 0.7 g NaOH in 25 mL of Type I water. Afterwards, 10.8 g $KNaC_4H_4O_6$ and 0.3 g $Na_2O_5S_2$ were added and mixed. Final volume was adjusted to 50 mL with addition of Type I water. The solution was stored in the refrigerator wrapped in aluminum foil to protect it from light.

2. **DNS reaction:** in a PCR tube, 24 μL of sample were mixed with 96 μL of DNS reagent. The mixture was heated at 95 °C for 5 min. Next, 20 μL of the heated mixture were transferred to a flat bottom 96-well plate filled with 80 μL of Type I water. Quantification of reducing sugars was conducted photometricaly by measuring light absorbance at 540 nm with a plate reader (CLARIOstar, BMG Labtech).

3. **D-fructose standard curve:** A D-fructose stock solution was prepared with a concentration of 100 mM in potassium phosphate buffer (200 mM, pH 7). This solution was used to make serial dilutions ranging between 0 and 50 mM. All stocks were prepared in triplicates. Reactions with DNS were performed as described above.

4.2.6.6 Analysis of carbohydrates using high performance liquid chromatography (HPLC)

Analytics for sucrose, G1P and fructose were performed using a HPLC Dionex Ultimate 3000 system equipped with autosampler (WPS 3000TRS), a column compartment (TCC3000RS) and a light scattering detector. The YMC Triart Diol Hilic column (100 x 2 mm, 1.9 μm, 12 nm) at 7 °C was used for separation by gradient elution (15-85 %) with 0.1 % formic acid pH 4.5, and 0.1 % formic acid in acetonitrile as the mobile phase at a flow of 0.45 mL/min. Samples were diluted in water/acetonitrile with the ratio of 3:7. Data were analyzed using Dionex Chromeleon software. Analysis of the samples was performed by CASCAT GmbH, Germany.

4.3 Results and discussion

4.3.1 Investigation of coat proteins and linkers as putative fusion partners for spore display

In order to find suitable anchoring proteins as fusion partners for the display of enzymes, four different coat proteins (CotB, CotG, CotY, CotZ) were chosen. The coat proteins CotB and CotG are components of the outer membrane of the spore coat while the coat proteins CotY and CotZ are components of the most outer shell of the spore, named crust [208, 99]. The coat proteins were evaluated in pairs by using fluorophores as reporters (Figure 4.5). To this end five genetic constructs were designed as shown in Table 4.3. Each genetic construct comprises two genes encoding for the anchoring proteins CotB, CotG, CotY and CotZ, C-terminally fused to the fluorescent proteins GFPmut2 and mKate2 respectively. The genes coding for the anchoring proteins and the fluorophores, are bridged by linkers in order to introduce a spacial distance between the fusion partners. Two different linkers were tested, one consists of glycerin/serine residues (GGGGGS)$_2$ that are not able to form a fixed structure, therefore is called flexible, whereas the other linker contains more polar amino acids (EAAAK)$_3$ that form an α-helix and therefore is called rigid [108]. All sequences were codon optimized for *B. subtilis*. The constructs were integrated into the *pksX* locus of the genome via homologous recombination and were expressed under the control of P*cotYZ* promoter. Constructs containing both *cotB* and *cotG* were integrated in strain Bs02007 where the native genes encoding for these two coat proteins were deleted. All other constructs were integrated in the genome of strain Bs02003. All generated strains shown in Table 4.3 are germination mutants.

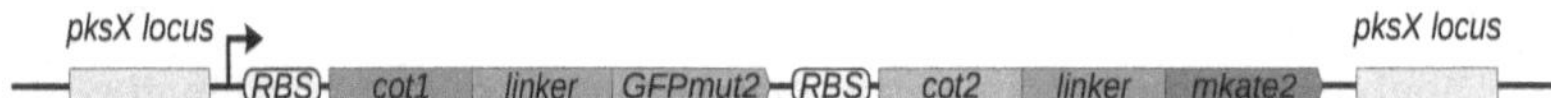

Figure 4.5: General scheme of genetic construct designed for the evaluation of coat proteins (cot1 = coat gene 1, cot2 = coat gene 2).

Table 4.3: Genetic constructs used for evaluation of the library components (cot1 = coat gene 1, cot2 = coat gene 2).

Strain	Cot1	Linker1	Reporter1	Cot2	Linker2	Reporter2
Bs17001	*cotG*	*flexible*	*GFPmut2*	*cotB*	*flexible*	*mKate2*
Bs17002	*cotG*	*rigid*	*GFPmut2*	*cotB*	*rigid*	*mKate2*
Bs17003	*cotY*	*flexible*	*GFPmut2*	*cotZ*	*flexible*	*mKate2*
Bs17004	*cotY*	*rigid*	*GFPmut2*	*cotZ*	*rigid*	*mKate2*
Bs17005	*cotG*	*rigid*	*GFPmut2*	*cotZ*	*flexible*	*mKate2*

The designed constructs were compared in regard to their fluorescence properties. As negative control, spores from strain Bs02003 were employed. For each construct different spore dilutions were prepared and two end point FI measurements were performed using a plate reader. For GFPmut2 excitation was done at 485 nm and emission was detected at 520 nm, whereas for mKate2 the respective wavelengths were 575 nm and 620 nm. Subsequently the measured fluorescence signals were normalized to the absorbance at 450 nm ($Abs_{450\,nm}$). Data were evaluated for statistical significance using one-way ANOVA and Tukey's test. The transformed signals are presented in Figure 4.6 while a detailed analysis of the statistical tests is available in Section 6.4.

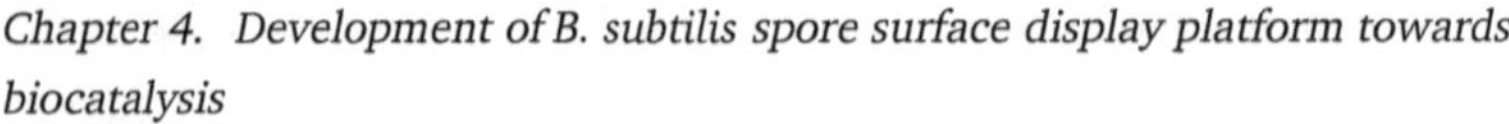

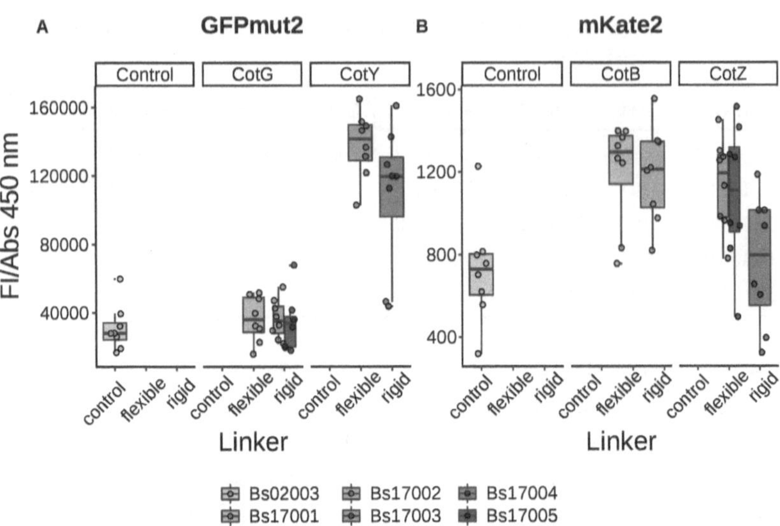

Figure 4.6: Evaluation of different anchoring proteins and linkers employed for spore display, using GFPmut2 (**A**) and mKate2 (**B**) as reporter genes. Strain Bs02003 is used as control. The normalized FI signals (FI/Abs$_{450\,nm}$) are depicted as dots, boxes represent Q1-Q3, lines represent median whereas the error bars show the standard deviation (n=8).

As shown in Figure 4.6A regarding GFPmut2, the highest fluorescence signal was obtained when the fluorophore was fused to the anchoring protein CotY whereas no statistically significant difference (p>0.05) was obtained between CotG fusion proteins and the control strain Bs02003, indicating a dysfunctional fusion or weak expression. In the case of mKate2 no statistically significant difference was obtained between CotB and CotZ fusions. Furthermore, the recorded signal for strain Bs17004 which expresses mKate2 as fusion to CotZ linked by rigid linker, was at similar levels as for the control strain. Results also revealed no significant difference (p>0.05) between flexible and rigid linker.

It is noteworthy that the fluorescence signal measured for the mKate2 constructs is 2 orders of magnitude lower than the signals measured for GFPmut2 indicating a possible weak expression, mRNA instability or low photostability. Similar observations were made by Norman et al. when expressing mKate2 in *B. subtilis* [140]. The authors overcame the problem of low expression by introducing a leader sequence from the overexpressed protein ComGA in the sequence of mKate2 [140].

Following, strain Bs17003 was transformed with the low copy plasmid p14035 which contains two homology regions to the *pksX* locus [131]. Via homologous recombination the genomic integrated sequences located in between the homology regions are copied on the plasmid vector thus rendering it replicative. This technology is called CopySwitch [131]. The generated strain was named Bs02034. Spores from this strain as well as from the parent strain Bs17003 and the control strain Bs02003 were compared in regard to fluorescence intensity (FI).

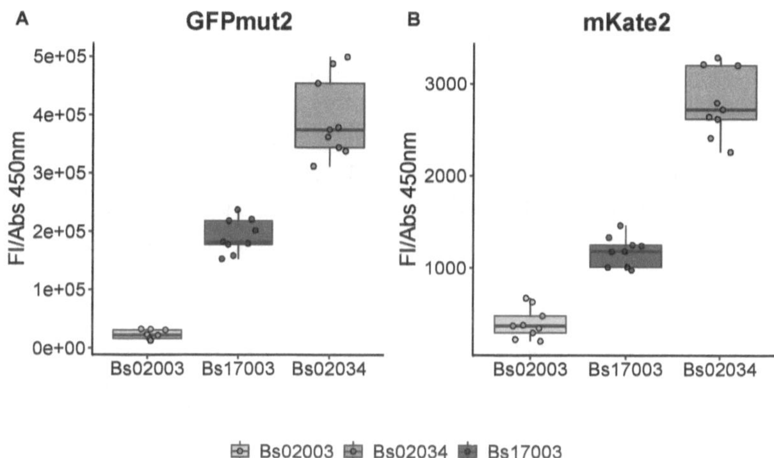

Figure 4.7: Box plot diagram comparing the normalized FI signals (FI/Abs$_{450\,nm}$) of spores from three different strains. Bs02003 is the control strain, Bs17003 harbors chromosomal integration of CotY-GFPmut2 and CotZ-mKate2, Bs02034 harbors low copy plasmid (p14035).

As shown in Figure 4.7 the obtained fluorescence signal for Bs02034 spores is 2 and 2.3 fold higher for GFPmut2 and mKate2 respectively compared to the values obtained for the spores from strain Bs17003. These results indicate that the CopySwitch method is applicable to spore display and is able to increase the protein load on the spore by expressing the gene of interest in multiple copies. These results are in line with the reported results from Nadler et al. who demonstrated an increase in the GFP production when applied the CopySwitch method in *B. subtilis* cells [131].

4.3.2 Spore display of *Cv*FAP for the production of biofuels

4.3.2.1 Effect of gene copy number in the activity of the spore displayed *Cv*FAP

For the display of *Cv*FAP on the spores, the coat protein CotY (18 kDA) was selected as the anchoring protein of the fusion. The selection of this protein was based on its location on the most outer part of the spore, the crust, and on our previous experiments suggesting better display efficiency compared to other coat proteins. The codon optimized gene encoding for *Cv*FAP was fused in frame to the C-terminal coding end of the *cotY* gene. The two genes were bridged by a glycerin/serine linker of 12 amino acids (GGGGGS)$_2$. The fusion was designed this way by taking into account the protein structure of *Cv*FAP, which revealed that the N-terminal site of the protein forms a long flexible helix (PDB:5NCC) [181]. At the C-terminal *Cv*FAP contained a 6 x His-tag to allow for its localization on the spore coat. Towards the production of efficient spore based catalysts, it was hypothesized that an increase in the gene copy number of the fusion would lead to higher enzyme concentration and activity as it was already shown in Figure 4.7. Therefore three strains were developed harboring different gene copy numbers of the fusion gene *cotY-Cv*FAP and its promoter. Strain Bs02035 harbors a single copy of the gene of interest integrated to the genome via homologous recombination, whereas strain Bs02037 and Bs02039 contain several copies of the fusion gene in low (p14035) and high copy (p02030) plasmids respectively. The low copy plasmid p14035 has a reported number of around 6 units/cell [131, 190] whereas the high copy plasmid p02030 which is derivative of pMSE3 has a reported number of copies around 300 units/cell [175]. The constructed strains are derivatives of the germination deficient strain Bs02003, therefore it was employed as negative control. The decarboxylation activity of the spore displayed *Cv*FAP was tested using palmitic acid in n-hexane as substrate for the production of pentadecane under blue light.

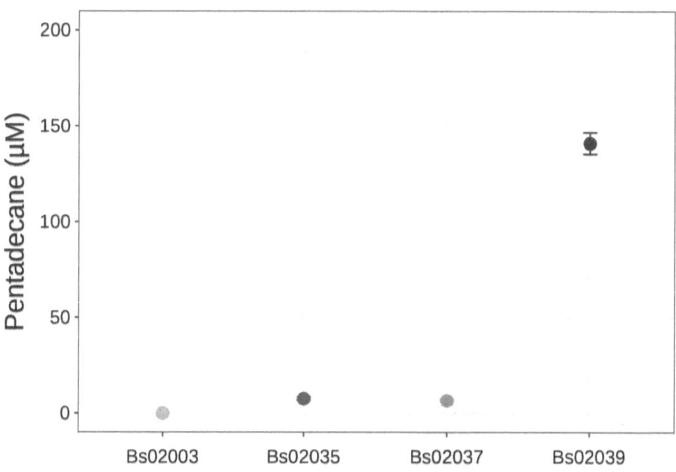

Figure 4.8: Gene dosage effect in the activity of spore displayed *Cv*FAP. Four different strains were tested: Bs02003 is the negative control, Bs02035 has a single copy (s.c) of the gene fusion *cotY-Cv*FAP, while strains Bs02037 and Bs02039 express the same fusion gene in a low copy (l.c.) and high copy plasmid (h.c.) respectively. Reaction conditions: 15 mM palmitic acid in n-hexane (50 % v/v), 5 mg/mL SDW, 30°C, 20 hours. Error bars indicate the standard deviation of four technical replicates.

As expected, the highest enzymatic activity was observed for strain Bs02039 which expresses the genetic fusion in the form of a high copy plasmid, whereas trace amounts of pentadecane were obtained for strains Bs02035 and Bs02037 (Figure 4.8). No enzymatic activity was monitored for the negative control strain Bs02003. These results demonstrate that up to 6 gene copies of the *cotY-Cv*FAP fusion, were not sufficient to increase the catalytic activity. Although the actual copy number of the plasmid used for Bs02039 was not specified in the current work, still these results indicate that the 19-fold higher activity can be associated with higher protein load on the spore surface stemming from the multiple gene copies. Xiaoman Xu et al. reported on similar results when N-*acetyl*-D-neuraminic acid aldolase was displayed on the *B. subtilis* spore surface using a low and a high copy vector [201].

Interestingly the overall reaction conversion did not exceed 1 %. This dramatically low yield could be associated with many factors such as weak protein expression, steric hindrance caused by the immobilization or even photoinactivation of the displayed enzyme. Lakavath et. al demonstrated that exposure of the catalyst to blue-light is directly linked to the formation of radical species, leading to irreversible inactivation of the enzyme [104]. Based on the better performance of Bs02039 all further experiments were conducted with this strain.

4.3.2.2 Effect of reaction conditions in the catalytic activity of the spore displayed *Cv*FAP

To gain deeper insight into the properties of the spore displayed *Cv*FAP, different factors with a putative effect in the catalytic activity toward decarboxylation of palmitic acid were tested. Photocatalytic reactions were performed in biphasic systems containing n-hexane.

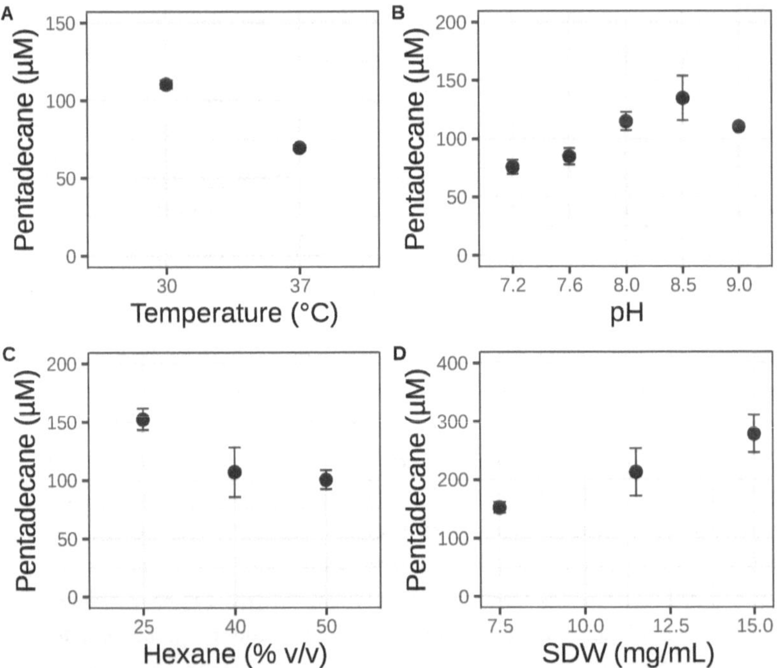

Figure 4.9: Enzymatic activity of the displayed *Cv*FAP (Bs02039) as a function of different reaction conditions. All reactions were performed using 15 mM palmitic acid in n-hexane as substrate and blue light illumination (68 μmol quanta m^{-2} s^{-1}). Incubation time was 20 hours. Reactions **B-D** were performed at 30 °C. Reaction conditions: **A.** 7 mg/mL SDW in Tris-HCl buffer (100 mM, pH 9.0), 50 % n-hexane. **B.** 7.5 mg/mL SDW in Tris-HCl buffer (100 mM), 50 % n-hexane **C.** 7.5 mg/mL SDW in Tris-HCl buffer (100 mM, pH 8.5). **D.** 7.5-15 mg/mL SDW in Tris-HCl buffer (100 mM, pH 8.5), 25 % n-hexane. Error bars indicate the standard deviation (n=3) [87].

As shown in Figure 4.9A an increase in reaction temperature from 30 °C to 37 °C was followed by a reduction in the catalytic activity of the enzyme. This effect could be explained by considering that *Cv*FAP originates from microalga which have optimal growth temperatures between 15 °C and 30 °C [154]. The pH profile (Figure 4.9B) results, show that maximum product formation is obtained at pH 8.5 whereas almost 50 % decrease was observed at pH 7.2. The obtained results are in accordance with the data presented for the free enzyme [181]. To investigate the organic solvent-tolerance of the catalyst, three different concentrations of n-hexane were employed. Figure 4.9C shows that product formation decreases in higher organic solvent concentrations suggesting the inactivation of the catalyst. Next, the influence of catalyst concentration was examined. As expected higher spore concentrations led accordingly to higher product formation (Figure 4.9D).

Despite the here presented efforts to optimize the reaction conditions for *Cv*FAP, the obtained pentadecane concentrations remained significantly low. In order to investigate if the low conversion could be associated with the low concentration of the co-factor FAD, spores were supplemented with FAD after purification. A small increase in the bioconversion was obtained, however addition of the FAD would significantly increase the cost of the process thus it was not applied in the following experiments (data not shown). By monitoring the product formation over time (Figure S7), we observed that the reaction was mainly completed within 3-4 hours, thus confirming previous findings reporting on the unstable nature of *Cv*FAP [104]. In order to investigate the possibility of enzyme deactivation due to high light intensity we tested two different light settings. It was observed that the enzymatic activity was decreased when using lower light intensity (Figure S9). These results are in line with previous findings from Bruder et al. where the authors obtained a significant increase in the hydrocarbon yields produced from *Cv*FAP when using higher intensity blue light lumen [15]. Hence it can be advocated that the utilization of a setting with higher light intensity could benefit the reaction yield.

4.3.2.3 Determination of kinetic constants for the spore displayed *Cv*FAP

To determine the kinetic constants of the spore displayed *Cv*FAP, photocatalytic reactions were performed in biphasic systems containing different concentrations of palmitic acid dissolved in n-hexane (25 % v/v). The substrate concentrations ranged from 0.5 to 10 mM. The kinetic parameters were determined following the classical Michaelis-Menten model. A non linear fit of the experimental data to the Michaelis-Menten model was performed using R studio [76]. Results of estimated Michaelis-Menten constant (K_M), maximum reaction velocity (V_{max}) and spore specific activity are shown in Table 4.4 .

Table 4.4: Kinetic parameters for the spore displayed *Cv*FAP. Reactions were performed using different concentrations of palmitic acid in 30 °C under blue light illumination (68 μmol quanta m^{-2} s^{-1}). Reactions contained 25% n-hexane and 6 mg/mL of SDW.

Strain	K_M (mM)	V_{max} (U[1]/L)	SDW specific activity (μM min^{-1} mg^{-1})
Bs02039	0.96 ± 0.2	1.36 ± 0.08	0.76

The K_M value was calculated 0.96 ± 0.2 mM. Only a few data are available regarding the kinetics of *Cv*FAP in its free form. Lakavath et al. recently determined the K_M for palmitic acid of 98.8 ± 53.3 μM [104]. Since the operating reaction conditions differed to the nature and amount of organic solvent used, no direct comparison can be made. However it is worth mentioning that immobilization of the catalyst frequently leads to higher K_M value compared to the free enzyme due to substrate diffusion and transport issues arising from the lower accessibility to the enzyme active center due to immobilization [170]. Although it is common to determine the *Kcat* constant when performing biocatalytic reactions in this case it was not possible due to the lack of information regarding the concentration of the catalyst immobilized on the spore surface. Therefore as a mean of comparison to other immobilization methods, in the present work SDW specific activity was defined which corresponds to enzymatic activity per mg of spore dry weight (SDW).

It should be noted that protein analysis experiments were carried out such as sodium dodecyl sulfate polyacrylamide gel electrophoresis (SDS-PAGE) and protein dot blots in order to quantify the amount of displayed enzyme. To our disappointment all efforts for protein isolation were unsuccessful leading us to the assumption that the enzyme is weakly expressed on the surface of the spores. Using fluorescence microscopy and immunolabelling it was possible to at least locate the protein of interest on the surface of the recombinant spores produced from strain Bs02035 thus confirming the presence of the enzyme on the spore

---

[1] 1 U = 1 μmol/min

coat (Figure S8).

4.3.2.4 Spore mediated biotransformation of triolein to hydrocarbons by employing a bienzymatic cascade

For the production of HCs from triglycerides (TAGs), as a proof of concept a one-pot bienzymatic catalysis was performed employing triolein as substrate. The enzymatic cascade comprises a lipase and the light inducible decarboxylase *Cv*FAP (Scheme 4.2) [75]. For the hydrolysis step two different variants of immobilized lipase from *Rhizopus oryzae* were tested. The one variant included the lipase displayed as a fusion protein on the surface of the spores (Bs95002) while the other variant was commercially available immobilized lipase (RO lipase). Furthermore the native hydrolytic activity of the spores was assessed by employing spores from strain Bs02003 without heterologous expression of lipase. The decarboxylation step was carried out by spores from strain Bs02039. To investigate the optimal conditions for the biotransformation, the cascade was carried out in one-step and two sequential steps respectively.

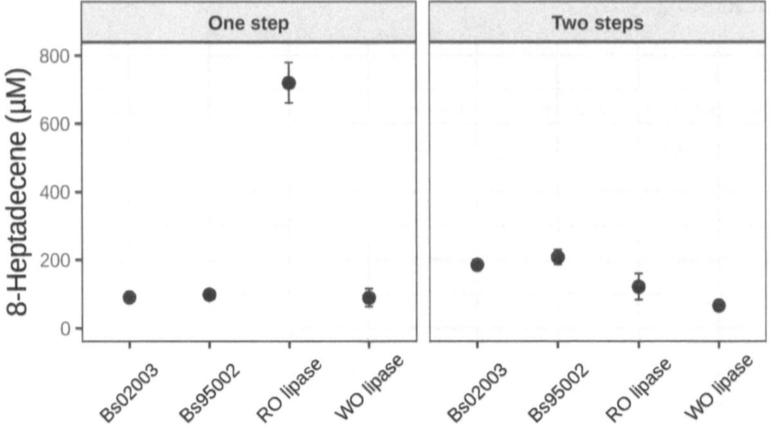

Figure 4.10: One pot bienzymatic conversion of triolein to 8-heptadecene (C17:1). The biotransformation was carried out in one and two step reactions respectively testing different lipases. One step reaction was carried out for 24 h. For the two step sequential catalytic reaction, Bs02039 spores were added after 48 h and the reaction was further carried out for 24 h. All reactions were performed in Tris-HCl buffer (100 mM, pH 8.5) using triolein as substrate and blue light illumination (68 μmol quanta m^{-2} s^{-1}). SDW concentration was 16 mg/mL. For the reactions where RO lipase was tested, 50 mg/mL of RO lipase were added. As control for the hydrolysis reaction, spores from strain Bs02003 were employed as well as reactions without addition of lipase were prepared (WO lipase). Data are mean values of triplicates. Error bars indicate standard deviation (n=3) [87].

As depicted in Figure 4.10, the maximum 8-heptadecene (C17:1) formation was obtained when the hydrolysis step was catalyzed by RO lipase in one-step biocatalysis, yielding a product concentration of approximately 0.7 mM which corresponds to roughly 3.5 % conversion. The titer levels were 3-fold higher than those attained with the two-step biotransformation indicating a better synergy between the displayed *Cv*FAP and the lipase in the one-step catalysis. Upon visual inspection the addition of spores in the reaction assisted in better emulsification of the individual components, thus featuring a more uniform distribution of substrates and catalysts. The emulsifying effect of spores and cells in biphasic systems has been investigated in previous studies [25, 38]. Furthermore, measurement of the pH after the hydrolysis step indicated a drop of the initial pH value from 8.5 to 7.2 which is below the optimal reaction pH for *Cv*FAP thus impairing its catalytic activity. No difference in the produced titers was obtained between the control spores (Bs02003) and the spores displaying *Rhizopus oryzae* lipase (Bs95002), indicating that the displayed enzyme was not active under the performed conditions. Interestingly the C17:1 titers measured in the two step biotransformation for the two strains, were elevated compared to the reaction where buffer was added as control for autohydrolysis of the substrate. These results demonstrate that the spores retain a certain hydrolytic activity. This activity could derive from the multiple enzymes comprising the spore coat. One such enzyme is LipC which has been identified as lipase [122]. However to date publications reporting on the substrate specificity and the role of the enzyme remain scarce [122].

4.3.2.5 Spore mediated production of drop in biofuels from natural oils employing *Cv*FAP

Microbial oils extracted from the oleaginous yeasts *C. oleaginosus* (Co. oil) and *Y. lipolytica* (Yali. oil) as well as olive oil were used as substrates for the production of HCs. Driven by the results for the triolein conversion, the biotransformation was carried out in one-pot one-step bienzymatic cascade consisting of the RO lipase and the spore displayed *Cv*FAP (Bs02039). As control, the reactions were performed without the addition of lipase. The produced titers of HCs and are shown in Figure 4.11 whereas the total amount of HCs obtained for each substrate is shown in Table 4.5.

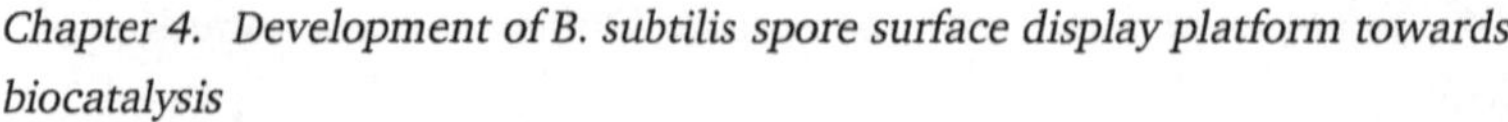

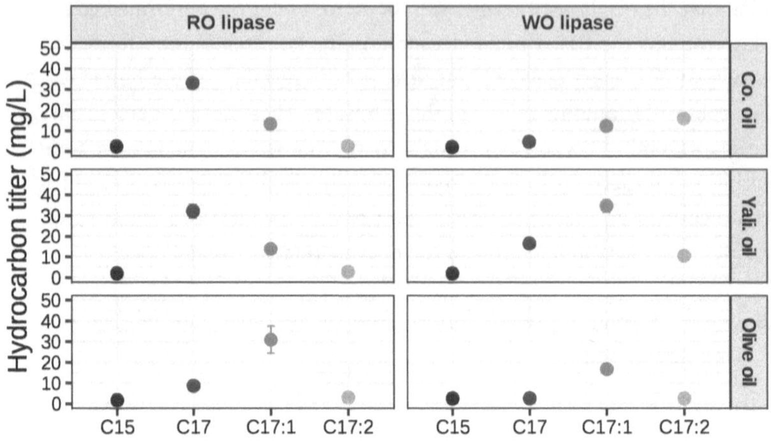

Figure 4.11: Biotransformation of natural oils (Co. oil, Yali oil, olive oil) to HCs catalyzed by the spore displayed *Cv*FAP (Bs02039). Profile of HCs produced with (RO) and without (WO) the addition of lipase. All reactions were performed in Tris-HCl buffer (100 mM, pH 8.5) using 25 μL of oil as substrate (2.5 % v/v) and blue light illumination (68 μmol quanta m^{-2} s^{-1}). SDW concentration was 16 mg/mL. For the reactions where lipase was tested, 50 mg/mL of RO lipase were added. Reaction time was 24 hours. Data are mean values of triplicates. Error bars indicate the standard deviation (n=3) [87].

Table 4.5: Titers of HCs produced from different oils with (RO) and without (WO) lipase.

Oil	Lipase	Titers (mg/L)
Co. oil	RO	51.5 ± 3.6
	WO	35.2 ± 5.2
Yali. oil	RO	50.8 ± 4.5
	WO	64.0 ± 5.6
Olive oil	RO	44.4 ± 8.9
	WO	24.5 ± 1.2

The selected oils have a fatty acid composition mainly made up of long-chain fatty acids [15, 123, 44]. As shown in Figure 4, in the reactions where microbial oils (Co. oil, Yali. oil) were used as lipid substrate, the addition of RO lipase increased production of heptadecane (C17), confirming the preference of *Cv*FAP for saturated fatty acids [75]. In contrast, without the addition of lipase (WO lipase) the hydrocarbon profile was changed in favor of unsaturated HCs, with

8-heptadecene (C17:1) and 6,9-heptadecadiene (C17:2) accounting in sum for more than 80 % and 70 % of the overall hydrocarbon fraction in each oil (Figure S11). This change was not observed when olive oil was used as substrate, with 8-heptadecene (C17:1) being the predominant hydrocarbon produced in both reactions. The sum of produced HCs (Table 4.5), indicates that in the cases of Co. oil and Olive oil, the presence of RO lipase resulted in only minor increase of the overall hydrocarbon titers while in the case of Yali. oil the highest titers (64.0 ± 5.6 mg/L) were obtained in the absence of lipase (WO lipase) corresponding to approximately 0.27 mM of HCs. These results suggest that the employed oils already contained a certain amount of free fatty acids most likely released during the extraction process or due to the hydrolytic activity of native lipases. To this end, the oils were analysed in regard to the free fatty acid and triglyceride content by employing HPLC. As shown in Figure S13, the Yali. oil contained the highest amount of free fatty acids compared to the two other oils which explains the increased hydrocarbon titers observed in the absence of the RO lipase. Furthermore we monitored the pH change before and after the biotransformation (Figure S14). A drop in the pH was observed in the presence of the lipase which is attributed to the release of free fatty acids in the reaction mixture. This pH change could compromise the activity of *Cv*FAP due to its lower stability in pH below 8.

Compared to other *in vitro* systems the here presented yield is significantly lower. For instance Ma et al. obtained up to 24 g/L of HCs when using soybean oil as substrate [117]. However since the amount of immobilized enzyme is not yet specified, no direct comparison can be made to systems that employ cell lysate. To sum up, the above results suggest that the rate limiting factor in the reaction is not the substrate availability but the stability and substrate specificity of *Cv*FAP.

4.3.3 Proof of concept: Spore display of sucrose phosphorylase for the production of glycosides

4.3.3.1 Display of sucrose phosphorylase from *Leuconostoc mesenteroides* (*Lm*SucP)

The sucrose phosphorylase from *L. mesenteroides* (*Lm*SucP) was displayed on the surface of the spores as a fusion to the C-terminal of the anchoring proteins CotY (Bs02028) and CotB (Bs02031). A 12 amino acids long linker was introduced in between the anchoring protein and the enzyme. The recombinant enzyme carries a 6 x His tag at the C-terminal. The gene sequence for *Lm*SucP was codon optimized for *B. subtilis* As control strain Bs02003 was employed. The release of α-Glc1P (G1P) and the consumption of sucrose were measured with HPLC as described under the Materials and Methods section.

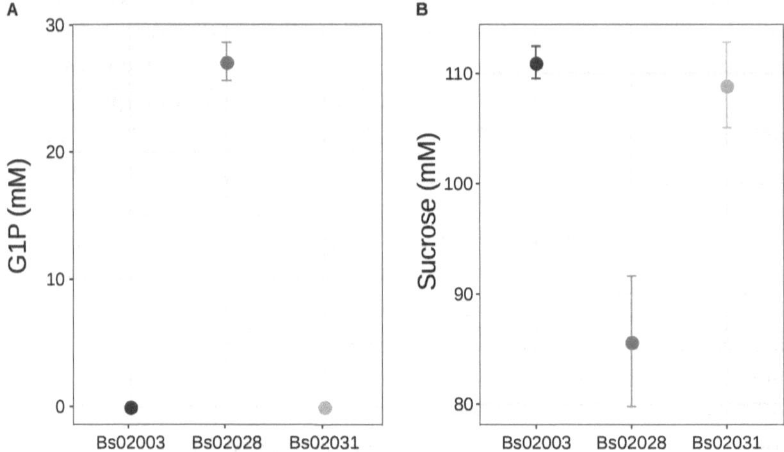

Figure 4.12: Phosphorolysis of sucrose to G1P and fructose from the spore displayed *Lm*SucP. Strain Bs02003 is used as control. Strain Bs02028 expresses SucP as a fusion to CotY whereas Bs02031 strain produces SucP as fusion to CotB. **A:** Formation of G1P for each strain. **B:** Consumption of sucrose. All reactions were performed in potassium phosphate buffer (100 mM, pH 6.5) using 100 mM sucrose as substrate and *B. subtilis* spores at a final OD $_{600\,nm}\sim$10. Reactions were performed at 30 °C for 24 h. Product formation and substrate consumption were measured by HPLC analysis. Error bars indicate the standard deviation (n=3).

To examine the activity of the spore displayed *Lm*SucP, samples from the reaction were analyzed in regard to the product formation and substrate consumption. Relatively low concentration of G1P was detected for strain Bs02028 which displays *Lm*SucP as a fusion to the crust protein CotY (Figure 4.12) whereas no activity was detected for the rest of the strains. The results for G1P formation are in consensus with the data obtained for the sucrose consumption, where a slight decrease in the concentration was observed only for strain Bs02028. It must be noted that due to the high sucrose concentration in the reaction, samples needed to be diluted prior to HPLC analysis, consequently driving the concentration of G1P to the lowest detection limit. The low enzymatic activity obtained may be associated with incorrect tertiary structure of the enzyme due to the fusion to the carrier protein or to insufficient loading of the recombinant protein on the spore coat. Furthermore, after 24 h of reaction *B. subtilis* cells were obtained, indicating that the presence of sugars in the mixture triggered germination and outgrowth of the recombinant spores. Hence this enzyme variant was excluded from further analysis.

4.3.3.2 Display of sucrose phosphorylase from *Bifidobacterium adolescentis* (*BaSucP*)

In view of the low enzymatic activity and spore germination obtained for the *Lm*SucP constructs, a sucrose phosphorylase variant (*Ba*SucP) with higher temperature optimum was tested. The codon optimized gene from *B. adolescentis* was heterologously expressed in *B. subtilis* as a fusion to the anchoring protein CotY under the control of the promoter P_{cotYZ}. Inspired by the results of gene dosage effect obtained for *Cv*FAP, the construct for *Ba*SucP was expressed in the form of a high copy plasmid. Recombinant spores were tested for their ability to convert sucrose and phosphate into G1P and fructose. Samples were analyzed using the DNS assay for detection of fructose [128].

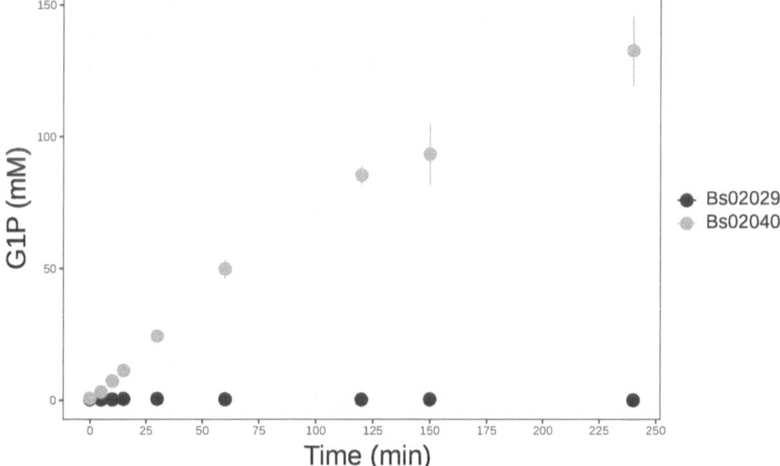

Figure 4.13: Phosphorolysis of sucrose to G1P over time from the spore displayed *Ba*SucP (Bs02040). Strain Bs02029 is used as control. The reaction was performed in potassium phosphate buffer (200 mM, pH 7.0) using 200 mM sucrose as substrate and *B. subtilis* spores at a final OD $_{600\,nm}$~0.7. Reactions were performed at 50 °C for 4 h. Quantification of G1P was carried out by specifying fructose formation with DNS assay. Data are mean values of triplicates. Error bars indicate the standard deviation (n=3).

The catalytic efficiency of the spore displayed *Ba*SucP was demonstrated by comparing its performance to spores from strain Bs2029 with no heterologous expression of sucrose phosphorylase at 50 °C (Figure 4.13). Within 4 hours of reaction the recombinant spores were able to convert up to 66 % of sucrose to equimolar amounts (132 mM) of fructose and G1P as indicated by the DNS assay corresponding to enzymatic activity of 0.732 U/mL. The here obtained activity is lower than the activity reported for the free enzyme produced from *B. subtilis* (5.3 U/mL) [198]. However it should be taken into consideration that a different assay was employed for quantification of G1P, thus no direct comparison should be made. The obtained results indicate that the enzyme was successfully dimerized during display. Notably for the present experiment only a small amount of spores was used ($OD_{600\ nm} \sim 0.7$) compared to the assay for *Lm*SucP. Moreover these results demonstrate that the displayed enzyme clearly outperforms the *Lm*SucP variant. Furthermore no signs of germinating spores were observed within the reaction time. It should be noted that utilization of DNS assay as detection method albeit being a fast quantification method has the drawback of being an indirect approach for quantification of G1P since the detection is based on the quantification of fructose (reducing sugar) present in the reaction. Furthermore it is advocated that glucose is also produced as result of the hydrolytic activity of sucrose phosphorylase. Hence absorbance values measured with the DNS assay are the transformed signals of fructose and glucose formed in the reaction.

4.3.3.3 Storage stability and reusability of spore displayed *Ba*SucP

The stability of an enzyme is particularly important for its applications. To test the long term storage stability of the immobilized enzyme, Bs02040 spores suspension in phosphate buffer (0.2 M, pH 7) were preserved at 4 °C for 102 days. Notably the displayed enzyme retained 88 % of its initial activity in this period. In order to more thoroughly investigate the storage stability of the enzyme more conditions including storage temperature and the physical form of the spores (liquid or powder) should be tested.

Table 4.6: Residual activity of spore displayed *Ba*SucP (Bs02040) after preservation at 4 °C for 102 days. Activity was determined using the DNS assay.

No. of days	Enzymatic activity (U^2/mL)	Residual activity (%)
1	0.739 ± 0.035	100
102	0.649 ± 0.017	88

One of the main advantages of using immobilized enzymes is the ease to retrieve them from the reaction mixture and reuse them several times. In this context, recombinant spores from strain Bs02040 were examined in regard to operational stability towards phosphorolysis of sucrose to G1P and fructose. Spores were reused up to 7 times in sequential reactions including a washing step before every cycle.

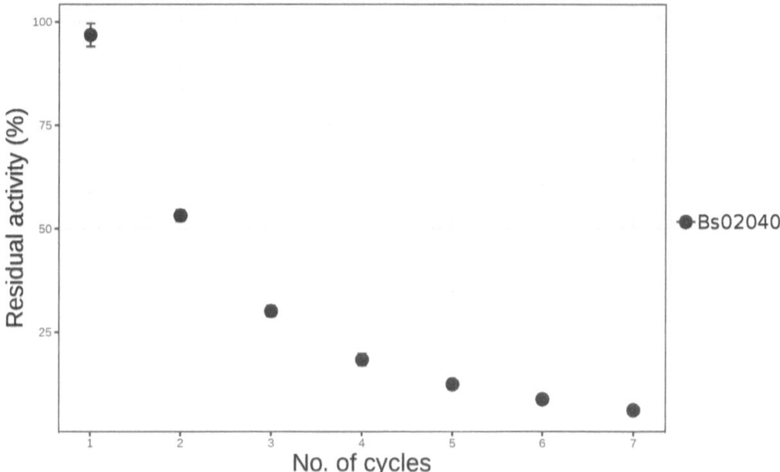

Figure 4.14: Operational stability of spore displayed *BaSucP* (Bs02040). Reactions were performed at 50 °C in potassium phosphate buffer (200 mM, pH 7.0) using 0.1 M sucrose as substrate and spores at a final OD $_{600\ nm}\sim$7. Quantification of fructose formation was carried out with DNS assay. Data are mean values of triplicates. Error bars indicate the standard deviation (n=3).

As shown in Figure 4.14 spores lost approximately 50 % of the initial activity already in the 2nd cycle while after six cycles the residual activity was reduced to 10 %. Compared to other immobilization methods where the authors immobilized the purified enzyme on Sepabeads and tested the continuous production of G1P for 250 h using a plug-flow reactor system, it seems that the spore displayed enzyme has lower stability [32]. However besides the degree of conversion one has to consider the economical cost of utilizing support materials for immobilization.

---

[2]1 U = 1 μmol/min

4.4   Conclusion and outlook

Bacterial spores have drawn immense interest as platforms for immobilization of enzymes by employing the fusion protein technology. In particular, spore surface display is an easy to perform and cost efficient method for presentation of proteins. The present book investigated the utilization of spore surface display in biocatalysis for the production of biofuels and glycosides. Initially a library of components was designed comprising of coat proteins utilized as carriers and two different linkers. The library was evaluated using the fluorescent proteins GFP-mut2 and mKate2 as reporters. The results indicated that the choice of carrier protein is important for the expression of the fusion while no effect was obtained among different linkers. However since the number of tested linkers was small, it is essential to construct a larger component library and evaluate linkers of different sizes. By recruiting replicative plasmids as expression vectors higher fluorescent signals were obtained indicating that the elevated gene copy number led to enhanced protein expression. These results suggest that the loading capacity of the spores surface can be further enhanced.

Following, it was shown for the first time that the spore display method can be employed for the production of biofuels by presenting a light inducible fatty acid decarboxylase (*Cv*FAP) coupled with a lipase in a bienzymatic assay. Several optimization steps were undertaken in order to enhance the catalytic efficiency of the displayed *Cv*FAP towards conversion of palmitic acid to pentadecane. Despite the efforts, the conversion did not exceed 2%. Many factors could contribute to this low efficiency, such as weak gene expression, photodeactivation of the catalyst, or low enzymatic activity due to steric hindrance caused by the immobilization. In order to elucidate the expression levels of the displayed enzyme, several approaches for protein extraction and isolation were tested. However quantification of the displayed enzyme was not achieved. These results further support the idea of weak expression of the enzyme. In the outlook phase of this work, the activity of spore displayed *Cv*FAP in a range of organic solvent systems should be tested especially due to the fact that this enzyme shows a catalytic preference for long chain fatty acids which have low solubility in aqueous systems. Finally spore display could serve as an ideal tool for protein engineering of *Cv*FAP and design of protein libraries, due to the direct genotype-phenotype coupling.

Spore display was further employed towards the production of G1P from the low cost substrate sucrose, by displaying sucrose phosphorylase. Two variants were tested with *Lm*SucP showing only trace amounts of product formation while *Ba*SucP could convert over 60 % of sucrose to G1P. Furthermore, the operational stability of the displayed *Ba*SucP was investigated by reusing the recombinant spores. These results are only a proof of concept study thus characterization of the displayed enzyme regarding its stability remain to be elucidated. In next steps, the spore displayed *Ba*SucP could be used for the continuous production of G1P or other glycosylated products using flow reactors.

5 Concluding remarks

The present doctoral book has provided new tools for the study of spores and has highlighted the potential of *B. subtilis* spores as biological platforms for immobilization of enzymes and biocatalysis. Most notably, the present study has illustrated that spore display is a simple to perform method that combines protein expression and purification in one step and has a broad applicability in different reaction concepts. Consequently, this method surpasses the tedious and time-money consuming immobilization methods that require the addition of support materials, thus considerably reducing the costs. Yet, many aspects of the display remain to be elucidated, such as the abundance of proteins on the spore coat, while tools for the accurate quantification of the displayed proteins need to be developed.

6 Supplementary material

6.1 Optimization of staining procedure for separation of cells and spores

To monitor the effect of staining in the viability of cells and spores, distinct samples of cells from sporulation deficient strain Bs02005 and spores from germinating strain Bs02002 were stained with SYBR1 and SYBR2 as mentioned above. Viability assessment was performed by sorting 210 events from each sample on an LB plate containing zeocin (20 mg/mL). Sorting was performed using a 100 μm microfluidics sorting chip for Sony SH800 (Sony), with the single cell three drops mode. Cells and spores were sorted on the plate according to the 384 well plate sort layout. Subsequently, plates were incubated overnight at 37 °C. For analysis, the relative number of colonies in respect to the total number of sorted events was evaluated.

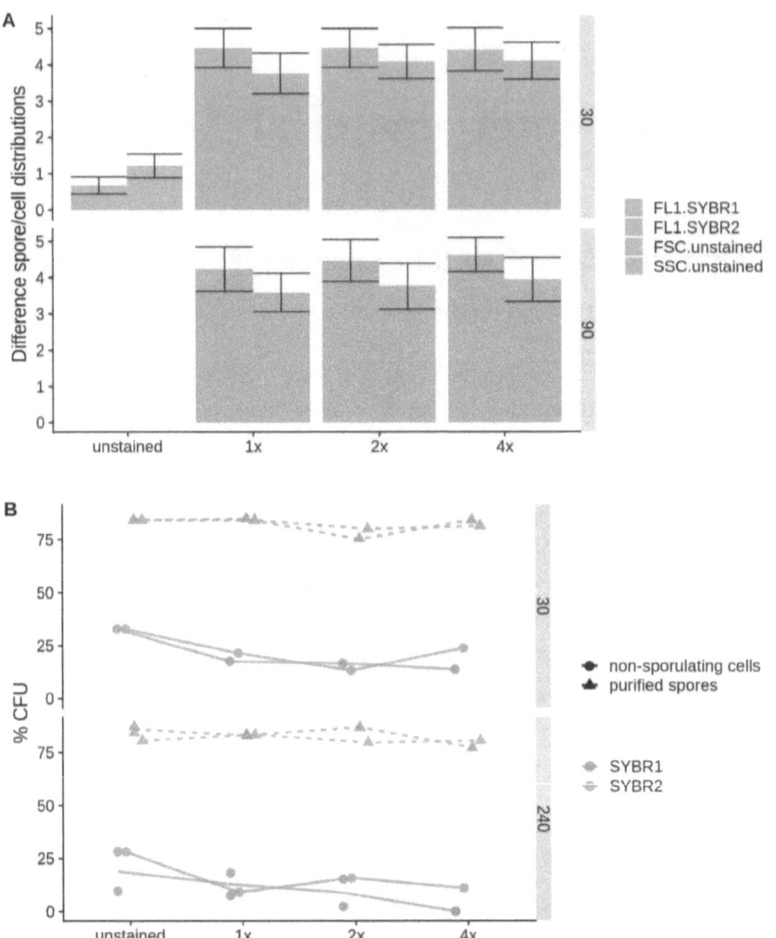

Figure S1: Optimization and evaluation of staining procedure. **A** shows variations in dye concentration and staining time. Differences of the distribution means as predicted by GMM with error bars showing pooled standard deviations are displayed. **B** shows cell survival after the same staining procedures. The percentage of viable cells is calculated as number of formed colonies in ratio to the total number of sorted events [86].

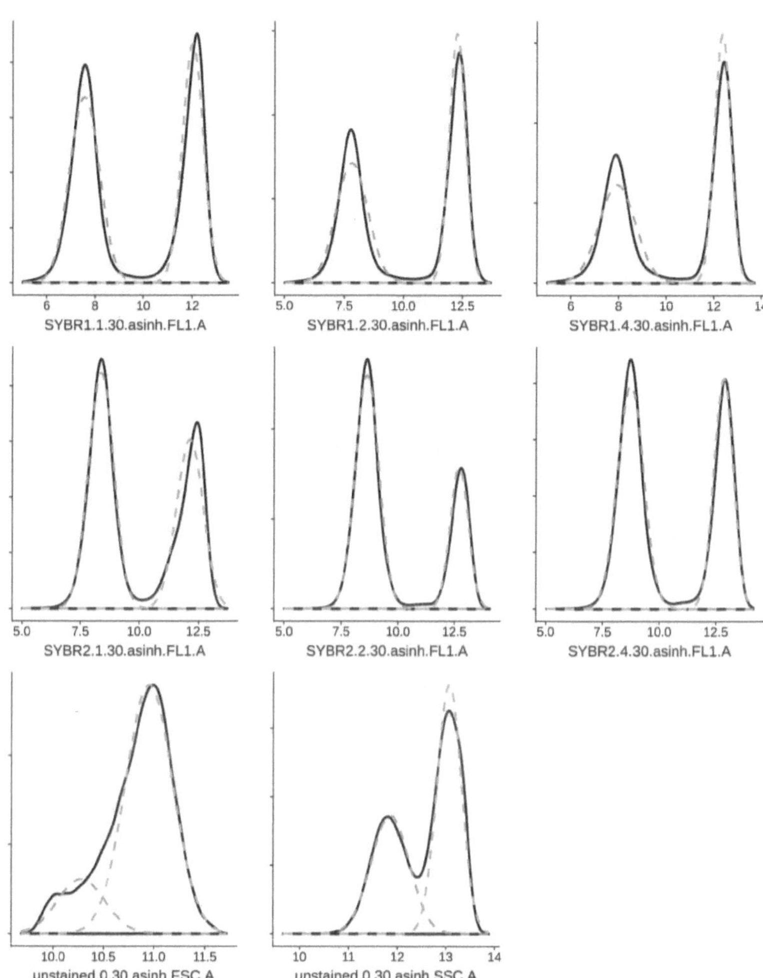

Figure S2: Separation of mixed cultures after 30 min of staining. Dashed lines indicate the predicted normal distributions of artificially mixed cell/spore samples as determined by GMM [86].

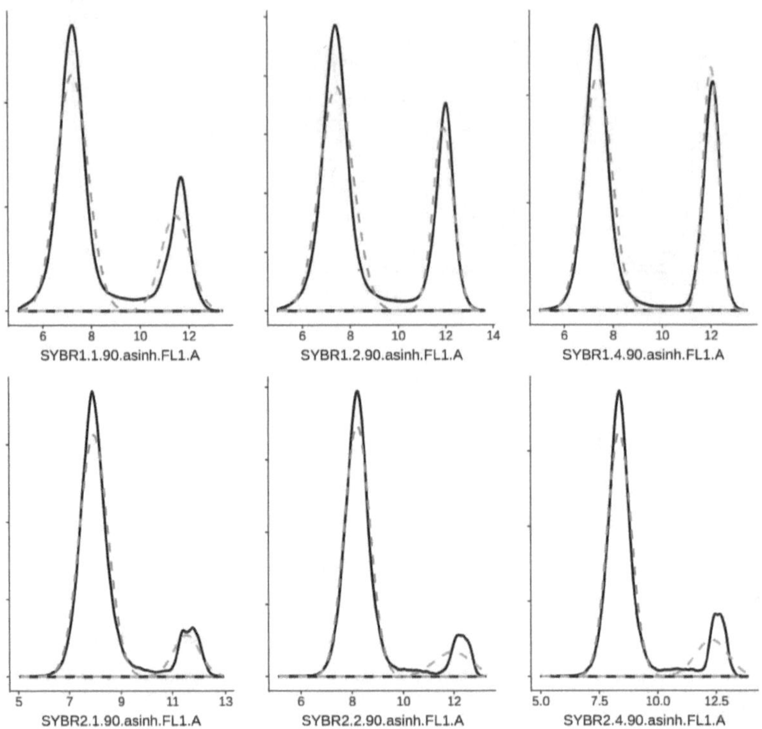

Figure S3: Separation of mixed cultures after 90 min of staining. Dashed lines indicate the predicted normal distributions of artificially mixed cell/spore samples as determined by GMM [86].

6.2   Cytometric analysis of Bs02003 and Bs02025.

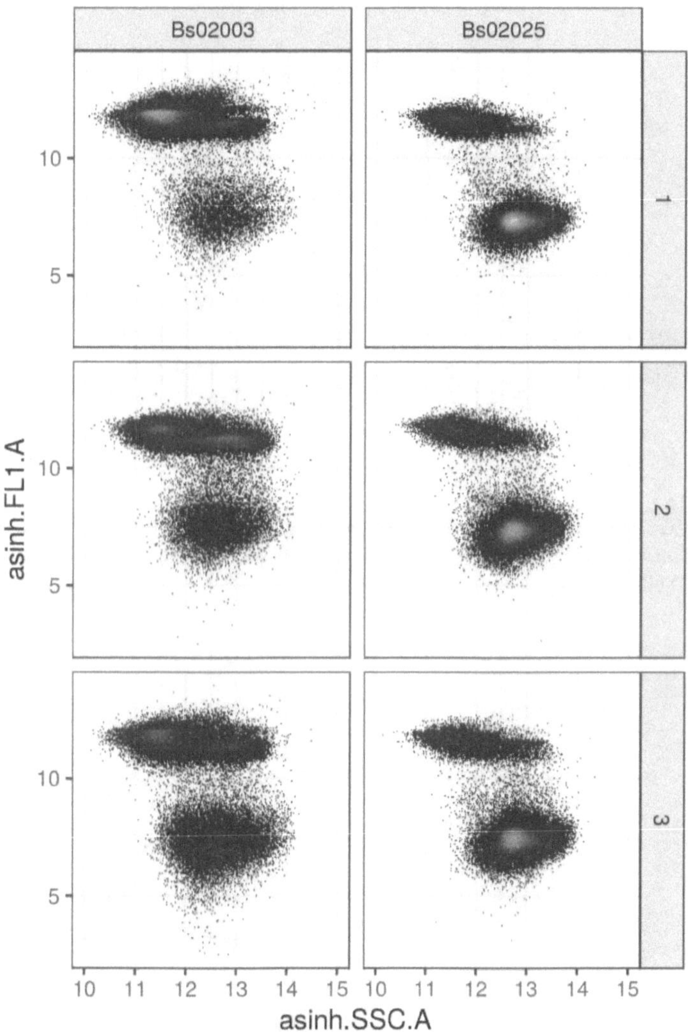

Figure S4: Scatter plots of raw data used in Figure 3.5B. Triplicates of Bs02003 and Bs02025 shown as scatter plots of SSC and SYBR1 (FL1) fluorescence [86].

6.3 Separation and quantification of spores using an automated robotic system.

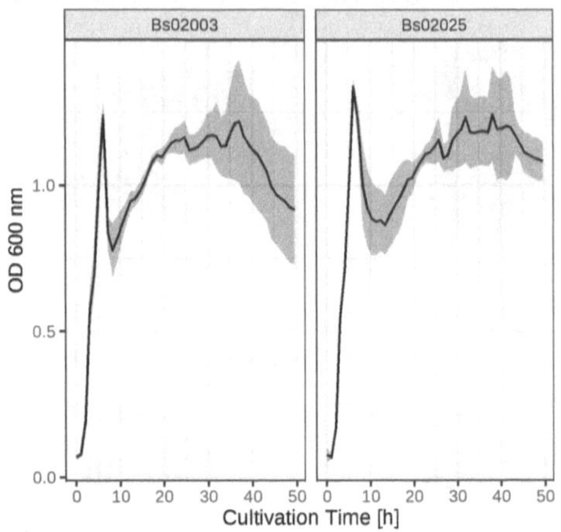

Figure S5: Growth curves of strains Bs02003 and Bs02025 cultivated in 2 x SG medium, employing a robotic platform. Gray shade indicates the standard deviation of four biological replicates.

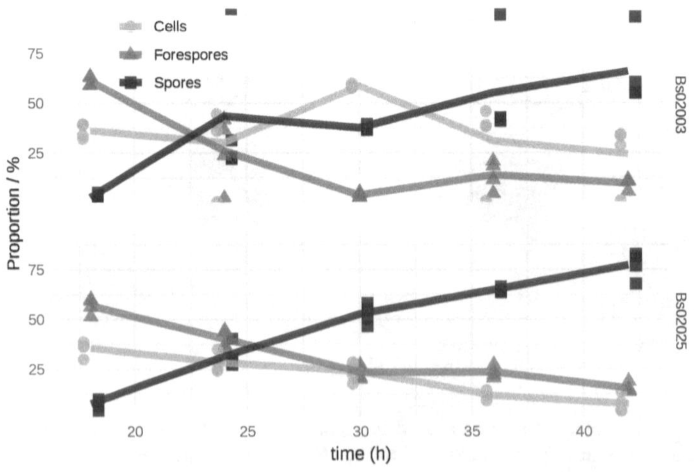

Figure S6: Sporulation dynamics of strains Bs02003 and Bs02025 in 2 x SG medium, employing a robotic platform.

6.4   Statistical analysis of data presented in Fig 4.6

```
#mKate
> ANOVA1<-lm(data1$norm.mKate2 ~ data1$Strain, data = data1)
> data.aov<-aov(ANOVA1)
> summary(data.aov)
             Df  Sum Sq Mean Sq F value  Pr(>F)
data1$Strain  5 1856158  371232   4.902 0.00127 **
Residuals    42 3180792   75733
---
Signif. codes:  0 '***' 0.001 '**' 0.01 '*' 0.05 '.' 0.1 ' ' 1
> TUKEY1<-TukeyHSD(data.aov)
> TUKEY1
  Tukey multiple comparisons of means
    95% family-wise confidence level

Fit: aov(formula = ANOVA1)

$`data1$Strain`
                       diff        lwr        upr      p adj
Bs17001-Bs02003  473.986799   63.221722 884.75188 0.0154447
Bs17002-Bs02003  464.599501   53.834424 875.36458 0.0185331
Bs17003-Bs02003  418.143051    7.377974 828.90813 0.0438940
Bs17004-Bs02003   41.206666 -369.558411 451.97174 0.9996521
Bs17005-Bs02003  363.064393  -47.700684 773.82947 0.1103111
Bs17002-Bs17001   -9.387298 -420.152375 401.37778 0.9999998
Bs17003-Bs17001  -55.843748 -466.608825 354.92133 0.9984811
Bs17004-Bs17001 -432.780133 -843.545209 -22.01506 0.0337075
Bs17005-Bs17001 -110.922406 -521.687483 299.84267 0.9648176
Bs17003-Bs17002  -46.456450 -457.221527 364.30863 0.9993757
Bs17004-Bs17002 -423.392835 -834.157912 -12.62776 0.0399617
Bs17005-Bs17002 -101.535108 -512.300185 309.22997 0.9759565
Bs17004-Bs17003 -376.936385 -787.701461  33.82869 0.0884787
Bs17005-Bs17003  -55.078658 -465.843735 355.68642 0.9985785
Bs17005-Bs17004  321.857727  -88.907350 732.62280 0.2016765


#GFP
> ANOVA2<-lm(data1$norm.GFP ~ data1$Strain, data = data1)
> data.aov2<-aov(ANOVA2)
> summary(data.aov2)
             Df    Sum Sq   Mean Sq F value   Pr(>F)
data1$Strain  5 8.819e+10 1.764e+10    36.1 3.65e-14 ***
Residuals    42 2.052e+10 4.886e+08
---
Signif. codes:  0 '***' 0.001 '**' 0.01 '*' 0.05 '.' 0.1 ' ' 1
> TUKEY2<-TukeyHSD(data.aov2)
> TUKEY2
  Tukey multiple comparisons of means
    95% family-wise confidence level

Fit: aov(formula = ANOVA2)

$`data1$Strain`
                        diff        lwr         upr      p adj
Bs17001-Bs02003    5241.36748  -27750.54  38233.280 0.9968093
Bs17002-Bs02003    5151.76996  -27840.14  38143.682 0.9970593
Bs17003-Bs02003  106895.04201   73903.13 139886.954 0.0000000
Bs17004-Bs02003   77837.17677   44845.26 110829.089 0.0000002
Bs17005-Bs02003    2704.72747  -30287.18  35696.640 0.9998708
Bs17002-Bs17001     -89.59752  -33081.51  32902.315 1.0000000
Bs17003-Bs17001  101653.67453   68661.76 134645.587 0.0000000
Bs17004-Bs17001   72595.80929   39603.90 105587.722 0.0000009
Bs17005-Bs17001   -2536.64000  -35528.55  30455.272 0.9999059
```

```
Bs17003-Bs17002   101743.27205     68751.36 134735.184 0.0000000
Bs17004-Bs17002    72685.40681     39693.49 105677.319 0.0000009
Bs17005-Bs17002    -2447.04249    -35438.95  30544.870 0.9999212
Bs17004-Bs17003   -29057.86523    -62049.78   3934.047 0.1125317
Bs17005-Bs17003  -104190.31453  -137182.23 -71198.402 0.0000000
Bs17005-Bs17004   -75132.44930  -108124.36 -42140.537 0.0000004
```

Linker comparison

```
#Anchor1_flex/rigid_norm.GFP
> ANOVA3<-lm(data6$norm.GFP ~ data6$Set1, data = data6)
> data.aov3<-aov(ANOVA3)
> summary(data.aov3)
            Df    Sum Sq   Mean Sq F value   Pr(>F)
data6$Set1   5 8.819e+10 1.764e+10    36.1 3.65e-14 ***
Residuals   42 2.052e+10 4.886e+08
---
Signif. codes:  0 '***' 0.001 '**' 0.01 '*' 0.05 '.' 0.1 ' ' 1
> TUKEY3<-TukeyHSD(data.aov3)
> TUKEY3
  Tukey multiple comparisons of means
    95% family-wise confidence level

Fit: aov(formula = ANOVA3)

$`data6$Set1`

                   diff              lwr         upr          p adj

Bs17002_rigid_  -89.59752      -33081.51     32902.315      1.0000000
CotG-
Bs17001_flexib
le_CotG

Bs17005_rigid_  -2536.64000    -35528.55     30455.272      0.9999059
CotG-
Bs17001_flexib
le_CotG

Bs17005_rigid_  -2447.04249    -35438.95     30544.870      0.9999212
CotG-
Bs17002_rigid_
CotG

Bs17004_rigid_  -29057.86523   -62049.78      3934.047      0.1125317
CotY-
Bs17003_flexib
le_CotY
```

```
#Anchor2_flex/rigid_norm.mkate2

> ANOVA4<-lm(data7$norm.mKate2 ~ data7$Set2, data = data7)
> data.aov4<-aov(ANOVA4)
> summary(data.aov4)
            Df  Sum Sq Mean Sq F value  Pr(>F)
data7$Set2   5 1856158  371232   4.902 0.00127 **
Residuals   42 3180792   75733
---
Signif. codes:  0 '***' 0.001 '**' 0.01 '*' 0.05 '.' 0.1 ' ' 1
> TUKEY4<-TukeyHSD(data.aov4)
> TUKEY4
  Tukey multiple comparisons of means
```

```
    95% family-wise confidence level

Fit: aov(formula = ANOVA4)

$`data7$Set2`
```

	diff	lwr	upr	p adj
Bs17002_rigid_CotB-Bs17001_flexible_CotB	-9.387298	-420.152375	401.37778	0.9999998
Bs17004_rigid_CotZ-Bs17003_flexible_CotZ	-376.936385	-787.701461	33.82869	0.0884787
Bs17005_flexible_CotZ-Bs17004_rigid_CotZ	321.857727	-88.907350	732.62280	0.2016765

6.5 Supplemental figures for spore displayed *Cv*FAP

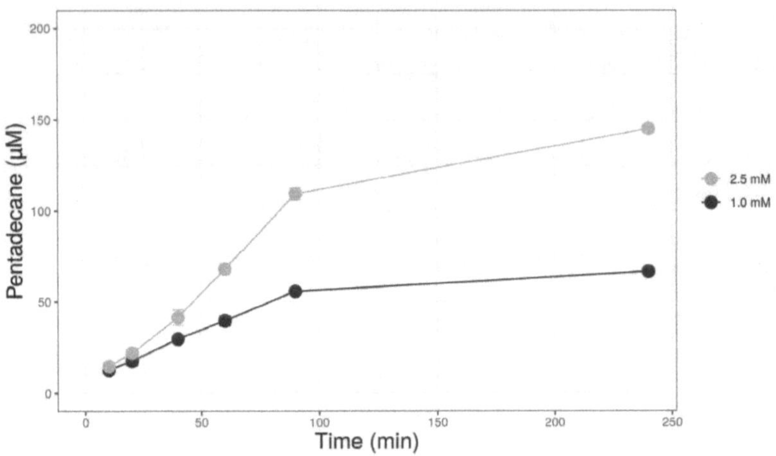

Figure S7: Decarboxylasion of palmitic acid to pentadecane over time catalysed by spore displayed *Cv*FAP. Reaction conditions: 1.0-2.5 mM palmitic acid, 6 mg/mL SDW (Bs02039), 25 % n-hexane, 30 °C under blue light illumination (68 μmol quanta m^{-2} s^{-1}) [87].

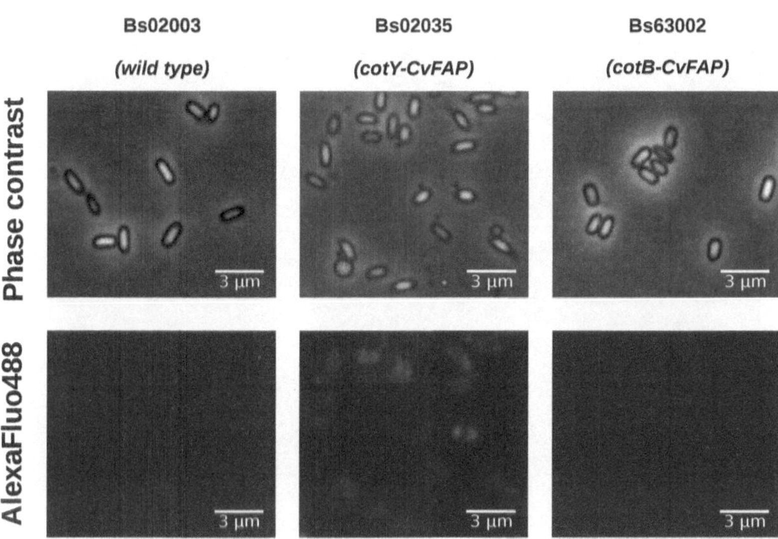

Figure S8: Immunodetection of *Cv*FAP on spores using AlexaFluo[488] anti-His tag antibody. Spores from Bs02003 were used as control. Bs020035: expresses *Cv*FAP as fusion to the anchoring protein CotY, while strain Bs63002 expresses *Cv*FAP as fusion to the anchoring protein CotB.

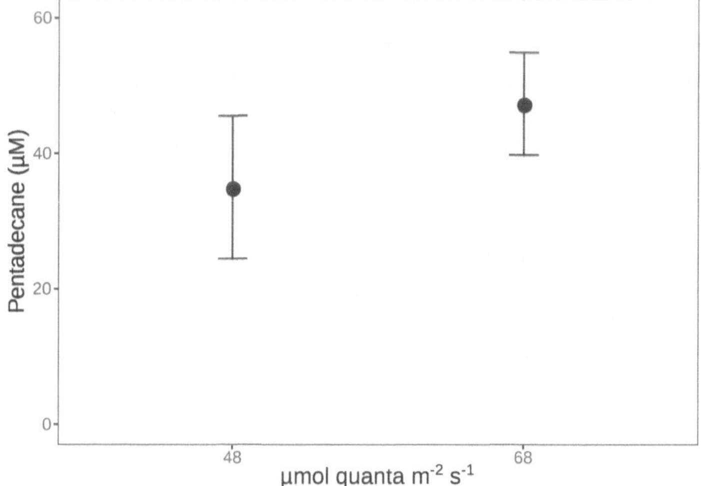

Figure S9: Effect of light on the activity of *Cv*FAP. Reaction conditions: 15 mM palmitic acid, spores (Bs02039, $OD_{600\ nm} \sim 40$), 30 °C, 70 rpm, 50 % n-hexane, Tris-HCl (100 mM, pH 9.0), blue light illumination [87].

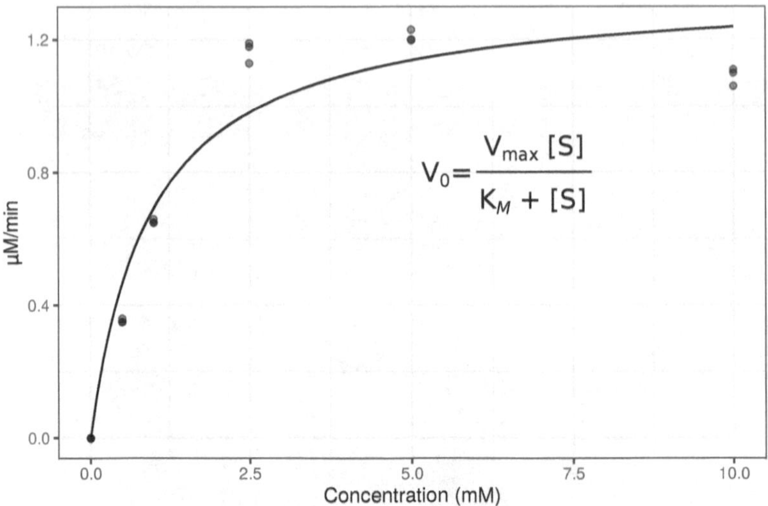

$$V_0 = \frac{V_{max}\,[S]}{K_M + [S]}$$

Figure S10: Michaelis-Menten curve of spore displayed *Cv*FAP (Bs02039) catalysing the decarboxylasion of palmitic acid. Reaction conditions: 0.5-10 mM palmitic acid, 6 mg/mL SDW, 30 °C, 70 rpm, 25 % n-hexane, Tris-HCl (100 mM, pH 9.0), blue light illumination (68 μmol quanta m^{-2} s^{-1}) [87].

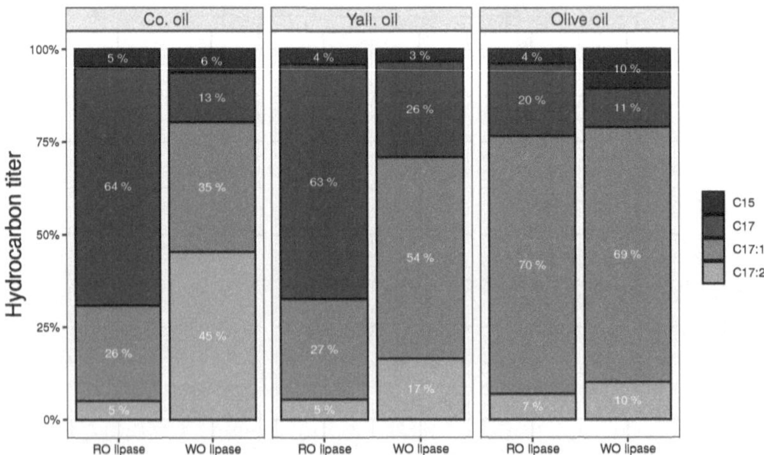

Figure S11: Composition of HC titers presented in Figure 4.11. RO lipase = *Rhizopus oryzae* lipase, WO lipase = without lipase [87].

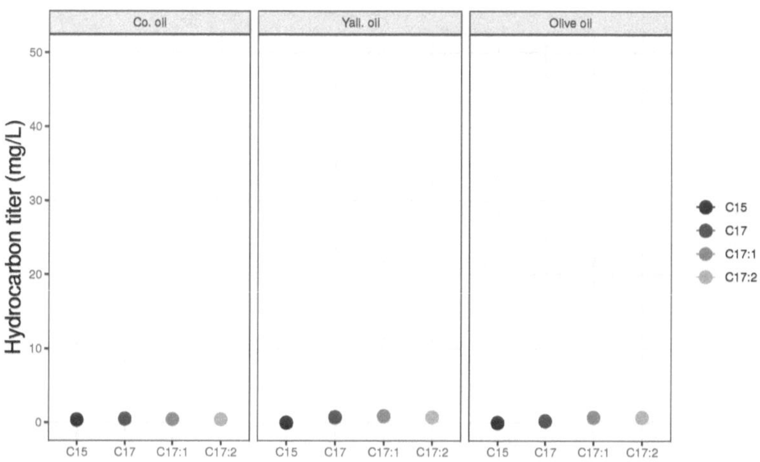

Figure S12: Hydrocarbon titers produced from Bs02003 spores using different oils as substrates [87].

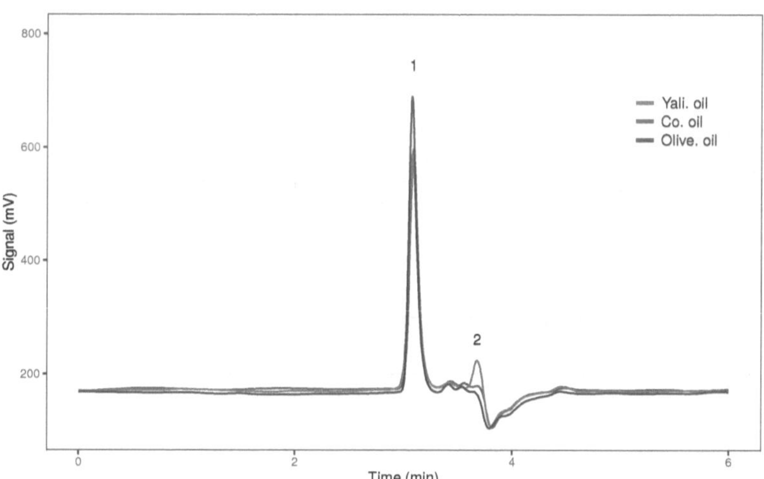

Figure S13: HPLC chromatograms of employed oils (Yali. oil, Co. oil, olive oil). The identified peaks correspond to: 1-triglycerides, 2-free fatty acids [87].

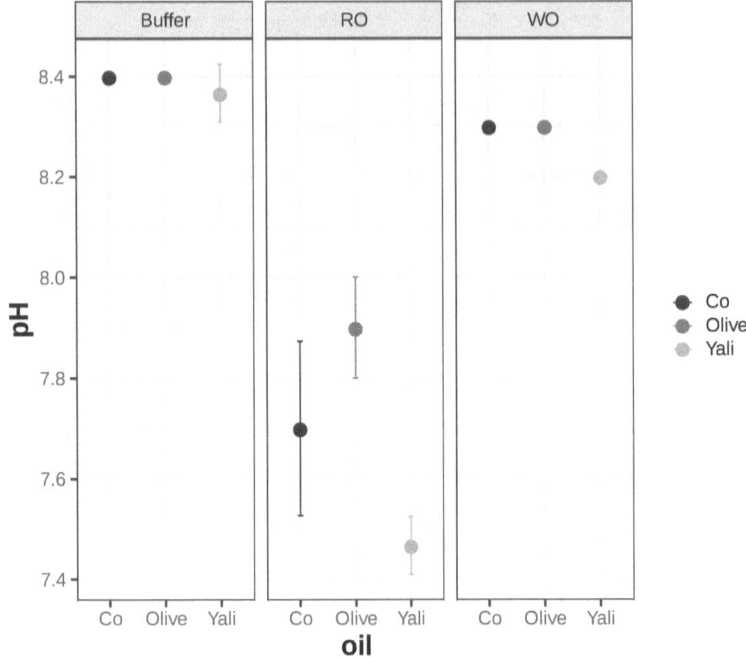

Figure S14: pH shift in biotransformation of oils (Yali. oil, Co. oil, olive oil). Reactions were performed as described in the material and methods section 4.2.4.4 (one-step biotransformation). Buffer: reactions contained only buffer and oil (2.5 % v/v), RO: reactions contained spores Bs02039 (16 mg/mL SDW), RO lipase (50 mg/mL) and oil (2.5 % v/v), WO: reactions contained spores Bs02039 (16 mg/mL SDW), buffer and oil (2.5 % v/v) [87].

6.6 DNA and protein sequences

(bold = coat protein, underlined = linker)

DNA sequence for cotY-<u>linker</u>-C*v*FAP-His-tag (p02027, p02030)

ATGAGCTGCGGAAAAACCCATGGCCGGCATGAGAACTGTGTATGCGATGCAGTGGAAA
AGATTTTAGCAGAGCAGGAGGCAGTTGAAGAACAGTGTCCGACTGGCTGCTATACCAA
CCTTTTAAACCCTACGATTGCTGGAAAAGACACAATTCCGTTTCTCGTTTTTGATAAAAA
AGGCGGATTGTTCTCCACATTCGGAAACGTAGGGGGGATTTGTGGATGATATGCAATGC
TTTGAATCCATTTTCTTCCGCGTCGAAAAATTATGCGATTGCTGTGCAACACTGTCTATT
TTACGCCCGGTCGATGTCAAAGGCGATACCTTAAGTGTTTGCCACCCTTGCGACCCGGA
TTTCTTCGGGCTAGAAAAAAACAGATTTCTGCATTGAAGTGGATCTCGGATGCTTCTGCG

CGATTCAGTGCCTGTCACCAGAGCTAGTTGACAGAACATCGCCTCACAAAGATAAAAAG
CATCATCACAATGGAGGTGGAGGTGGTGGAAGTGGTGGAGGTGGTGGAAGTATGAGAG
CGAGCGCAGTGGAAGATATCAGAAAGGTTCTTAGCGATAGCAGCTCACCGGTGGCGGGC
CAAAAATATGATTATATTCTGGTTGGCGGCGGCACAGCAGCATGCGTCCTGGCGAACAG
ACTGAGCGCGGATGGCTCAAAACGCGTTCTGGTCCTGGAAGCGGGACCGGATAATACAA
GCCGCGATGTTAAAATTCCGGCGGCAATTACAAGACTGTTTAGATCACCGCTCGATTGG
AATTTATTTAGCGAGCTGCAAGAACAACTGGCGGAAAGACAAATTTATATGGCTAGAGG
CAGACTGTTAGGCGGCAGCAGCGCAACAAATGCCACACTGTATCATAGAGGAGCGGCGG
GCGATTATGATGCGTGGGGCGTCGAAGGATGGTCTTCAGAAGATGTCCTGTCTTGGTTT
GTGCAAGCGGAAACAAATGCAGATTTTGGGCCTGGCGCGTATCATGGCAGCGGTGGACC
GATGCGAGTGGAAAACCCGAGATATACAAATAAACAACTGCATACAGCGTTTTTCAAAGC
AGCGGAAGAAGTAGGCTTAACACCGAACTCCGATTTCAACGACTGGTCACACGATCACG
CCGGCTACGGGACATTTCAGGTCATGCAGGATAAGGGAACACGGGCTGATATGTATCGT
CAGTATCTCAAACCGGTGCTTGGACGTCGAAATCTCCAGGTGTTGACGGGTGCTGCGGT
CACGAAAGTCAACATCGACCAGGCTGCAGGCAAAGCTCAGGCGCTGGGAGTTGAATTTT
CAACAGATGGACCGACAGGAGAAAGACTGAGCGCAGAATTAGCACCGGGCGGCGAAGT
TATTATGTGCGCGGGCGCGGTGCATACACCGTTTCTGCTGAAACATTCTGGCGTTGGCC
CGTCAGCAGAACTGAAAGAATTTGGAATTCCGGTTGTTTCAAATCTTGCGGGAGTTGGC
CAAAATCTGCAAGATCAGCCGGCATGTCTGACAGCGGCGCCGGTTAAAGAAAAATATGA
TGGCATTGCAATTAGCGATCATATTTATAATGAAAAAGGACAAATTAGAAAACGCGCAAT
TGCAAGCTATCTGCTGGGCGGCCGCGGCGGCCTGACATCCACAGGCTGCGATAGAGGCG
CGTTTGTGAGAACAGCTGGACAAGCGCTGCCGGATCTGCAGGTGAGATTTGTTCCGGGC
ATGGCGCTTGATCCGGATGGAGTCAGCACATATGTTAGATTTGCAAAATTTCAATCACAA
GGCCTGAAATGGCCGTCAGGAATTACAATGCAACTTATTGCGTGCCGCCCGCAGAGCAC
AGGCTCAGTGGGACTTAAATCTGCAGATCCGTTTGCGCCGCCGAAATTATCACCGGGCT
ATCTGACAGATAAAGATGGCGCCGATCTTGCAACACTTAGAAAAGGCATTCATTGGGCA
AGAGATGTGGCGAGAAGCAGCGCGCTTAGCGAATATCTTGATGGCGAATTATTTCCGGG
CAGCGGCGTTGTGTCAGATGATCAAATTGATGAATATATTAGAAGATCAATTCATAGCTC
AAATGCGATTACAGGAACATGCAAAATGGGCAATGCGGGCGATAGCTCAAGCGTTGTCG
ATAATCAACTGAGAGTGCATGGCGTTGAAGGACTGAGAGTTGTTGATGCAAGCGTTGTG
CCGAAAATTCCGGGCGGCCAAACAGGCGCGCCGGTTGTTATGATTGCGGAAAGAGCGGC
AGCGCTTCTGACAGGCAAAGCAACAATTGGCGCGAGCGCAGCAGCACCGGCGACAGTTG
CAGCACATCATCATCATCATCACTAA

Protein sequence for cotY-<u>linker</u>-C*v*FAP-His-tag

MSCGKTHGRHENCVCDAVEKILAEQEAVEEQCPTGCYTNLLNPTIAGKDTIPFLVFDKKG
GLFSTFGNVGGFVDDMQCFESIFFRVEKLCDCCATLSILRPVDVKGDTLSVCHPCDPDFFG
LEKTDFCIEVDLGCFCAIQCLSPELVDRTSPHKDKK<u>HHHNGGGGGSGGGGGS</u>MRASAVE
DIRKVLSDSSSPVAGQKYDYILVGGGTAACVLANRLSADGSKRVLVLEAGPDNTSRDVKIPAA

ITRLFRSPLDWNLFSELQEQLAERQIYMARGRLLGGSSATNATLYHRGAAGDYDAWGVEGW
SSEDVLSWFVQAETNADFGPGAYHGSGGPMRVENPRYTNKQLHTAFFKAAEEVGLTPNSDF
NDWSHDHAGYGTFQVMQDKGTRADMYRQYLKPVLGRRNLQVLTGAAVTKVNIDQAAGKA
QALGVEFSTDGPTGERLSAELAPGGEVIMCAGAVHTPFLLKHSGVGPSAELKEFGIPVVSNLA
GVGQNLQDQPACLTAAPVKEKYDGIAISDHIYNEKGQIRKRAIASYLLGGRGGLTSTGCDRG
AFVRTAGQALPDLQVRFVPGMALDPDGVSTYVRFAKFQSQGLKWPSGITMQLIACRPQSTG
SVGLKSADPFAPPKLSPGYLTDKDGADLATLRKGIHWARDVARSSALSEYLDGELFPGSGVVS
DDQIDEYIRRSIHSSNAITGTCKMGNAGDSSSVVDNQLRVHGVEGLRVVDASVVPKIPGGQT
GAPVVMIAERAAALLTGKATIGASAAAPATVAAHHHHHH*

DNA sequence for cotY-<u>linker</u>-*Lm*SucP-His-tag (p02026)

ATGAGCTGCGGAAAAACCCATGGCCGGCATGAGAACTGTGTATGCGATGCAGTGGAAA
AGATTTTAGCAGAGCAGGAGGCAGTTGAAGAACAGTGTCCGACTGGCTGCTATACCAA
CCTTTTAAACCCTACGATTGCTGGAAAAGACACAATTCCGTTTCTCGTTTTTGATAAAAA
AGGCGGATTGTTCTCCACATTCGGAAACGTAGGGGGGATTTGTGGATGATATGCAATGC
TTTGAATCCATTTTCTTCCGCGTCGAAAAATTATGCGATTGCTGTGCAACACTGTCTAT
TTTACGCCCGGTCGATGTCAAAGGCGATACCTTAAGTGTTTGCCACCCTTGCGACCCGG
ATTTCTTCGGGCTAGAAAAAACAGATTTCTGCATTGAAGTGGATCTCGGATGCTTCTGC
GCGATTCAGTGCCTGTCACCAGAGCTAGTTGACAGAACATCGCCTCACAAAGATAAAAA
GCATCATCACAATGGA<u>GGTGGAGGTGGTGGAAGTGGTGGAGGTGGTGGAGGT</u>ATGGAA
ATTCAAAATAAAGCGATGCTGATTACGTATGCCGATTCACTTGGAAAAAACCTGAAAGAT
GTTCATCAGGTTCTGAAAGAAGATATTGGCGATGCAATTGGCGGCGTGCACCTGCTACC
GTTTTTCCCGAGCACAGGGGATCGCGGCTTTGCGCCGGCGGACTATACACGCGTTGATG
CAGCGTTTGGCGACTGGGCAGACGTCGAGGCACTGGGCGAAGAGTACTACCTGATGTTT
GATTTTATGATAAACCATATCTCCAGAGAGTCTGTGATGTATCAGGATTTTAAAAAGAAT
CATGATGATTCAAAATATAAAGATTTCTTTATTCGTTGGGAAAAATTTTGGGCTAAAGCT
GGAGAAAATAGACCGACACAAGCAGATGTTGATCTGATTTATAAAAGAAAAGATAAAGC
TCCGACACAAGAAATTACATTTGATGATGGAACAACAGAAAATTTATGGAATACATTTGG
AGAAGAACAAATTGATATTGATGTTAATTCTGCTATTGCTAAGGAGTTCATCAAAACAAC
GCTCGAGGATATGGTCAAACATGGCGCGAACCTGATCCGCCTGGACGCGTTCGCGTACG
CCGTGAAAAAAGTTGACACAAACGACTTTTTCGTAGAACCTGAAATCTGGGATACACTGA
ACGAAGTCCGCGAAATCCTGACCCCGCTGAAAGCGGAGATTCTGCCTGAAATTCATGAA
CATTACAGCATCCCTAAGAAGATCAATGATCATGGCTATTTCACGTACGATTTCGCGCTT
CCGATGACGACGCTTTATACACTGTATAGCGGCAAAACCAACCAGCTTGCTAAATGGTTA
AAAATGTCTCCGATGAAGCAGTTCACGACCCTGGACACCCATGACGGGATCGGAGTTGT
CGACGCGCGCGACATTTTGACAGACGACGAGATTGATTATGCGTCTGAACAGCTGTATA
AGGTAGGCGCGAACGTCAAGAAGACGTATAGCAGCGCAAGCTACAACAACCTGGATATC
TATCAGATCAACTCCACATATTATTCCGCGCTCGGCAACGACGACGCGGCCTATCTGCTT
TCACGCGTGTTTCAGGTGTTTGCGCCGGGCATTCCGCAGATCTATTATGTGGGGCTGCT
GGCCGGAGAGAACGATATTGCGCTGCTTGAATCCACCAAGGAAGGCCGCAACATCAACC

GGCATTACTACACACGGGAAGAAGTTAAATCTGAAGTGAAACGGCCGGTGGTCGCAAAT
CTCCTGAAACTTCTGAGCTGGCGCAACGAAAGTCCGGCCTTTGACCTGGCGGGATCAAT
CACGGTGGACACGCCGACGGATACGACGATTGTGGTGACGAGACAGGATGAAAATGGCC
AAAATAAAGCGGTGCTGACAGCGGATGCGGCCAACAAAACATTTGAAATCGTGGAAAAC
GGCCAAACAGTCATGAGCAGCGATAACCTGACACAGAACCATCATCATCATCATCACTAA

Protein sequence for cotY-<u>linker</u>-*Lm*SucP-His-tag

**MSCGKTHGRHENCVCDAVEKILAEQEAVEEQCPTGCYTNLLNPTIAGKDTIPFLVFDKKG
GLFSTFGNVGGFVDDMQCFESIFFRVEKLCDCCATLSILRPVDVKGDTLSVCHPCDPDFFG
LEKTDFCIEVDLGCFCAIQCLSPELVDRTSPHKDKKHHHN<u>GGGGGSGGGGGG</u>MEIQNK
AMLITYADSLGKNLKDVHQVLKEDIGDAIGGVHLLPFFPSTGDRGFAPADYTRVDAAFGDWA
DVEALGEEYYLMFDFMINHISRESVMYQDFKKNHDDSKYKDFFIRWEKFWAKAGENRPTQA
DVDLIYKRKDKAPTQEITFDDGTTENLWNTFGEEQIDIDVNSAIAKEFIKTTLEDMVKHGANL
IRLDAFAYAVKKVDTNDFFVEPEIWDTLNEVREILTPLKAEILPEIHEHYSIPKKINDHGYFTYD
FALPMTTLYTLYSGKTNQLAKWLKMSPMKQFTTLDTHDGIGVVDARDILTDDEIDYASEQLY
KVGANVKKTYSSPASYNNLDIYQINSTYYSALGNDDAAYLLSRVFQVFAPGIPQIYYVGLLAGE
NDIALLESTKEGRNINRHYYTREEVKSEVKRPVVANLLKLLSWRNESPAFDLAGSITVDTPTD
TTIVVTRQDENGQNKAVLTADAANKTFEIVENGQTVMSSDNLTQNHHHHHH***

DNA sequence for cotB-<u>linker</u>-*Lm*SucP-His-tag (p02025)

ATGAGCAAGAGGAGAATGAAATATCATTCAAATAATGAAATATCGTATTATAACTTTTT
GCACTCAATGAAAGATAAAATTGTTACTGTATATCGTGGAGGTCCGGAATCTAAAAAAG
GAAAATTAACAGCTGTAAAATCAGATTATATAGCTTTACAAGCTGAAAAAAAAATAATT
TATTATCAGTTGGAGCATGTGAAAAGTATTACTGAGGATACCAATAATAGCACCACAAC
AATTGAGACTGAGGAAATGCTCGATGCTGATGATTTTCATAGCTTAATCGGACATTTAA
TAAACCAATCAGTTCAATTTAACCAAGGGGGTCCGGAATCTAAAAAAGGAAGATTGGTC
TGGCTGGGAGATGATTACGCTGCGTTAAACACAAATGAGGATGGGGTAGTGTATTTTA
ATATCCATCACATCAAAAGTATAAGTAAACACGAGCCTGATTTGAAAATAGAAGAGCAG
ACGCCAGTTGGAGTTTTGGAAGCTGATGATTTAAGCGAGGTTTTTAAGAGTCTGACTCA
TAAATGGGTTTCAATTAATCGTGGAGGTCCGGAAGCCATTGAGGGTATCCTTGTAGAT
AATGCCGACGGCCATTATACTATAGTGAAAAATCAAGAGGTGCTTCGCATCTATCCTTT
TCACATAAAAAGCATCAGCTTAGGTCCAAAAGGGTCGTACAAAAAAGAGGATCAAAAA
AATGAACAAAACCAGGAAGACAATAATGATAAGGACAGCAATTCGTTCATTTCTTCAAA
ATCATATAGCTCATCAAAATCATCTAAACGATCACTAAAATCTTCAGATGATCAATCATC
CAAATCTGGTCGTTCGTCACGTTCAAAAAGTTCTTCAAAATCATCTAAACGATCACTAA
AATCTTCGGATTATCAATCATCCAAATCTGGCCGTTCGTCACGTTCAAAAAGTTCTTCA
AAATCATCTAAACGATCATTAAAAATCTTCAGATTATCAATCATCAAAATCATCTAAACGA
TCACCAAGATCTTCAGATTATCAATCATCAAGATCACCAGGCTATTCAAGTTCAATAAA
AAGTTCAGGAAAACAAAAGGAAGATTATAGCTATGAAACGATTGTCAGAACGATAGAC
TATCACTGGAAACGTAAATTT<u>GGTGGAGGTGGTGGAAGTGGTGGAGGTGGTGGAGGTA</u>

Chapter 6. Supplementary material

TGGAAATTCAAAATAAAGCGATGCTGATTACGTATGCCGATTCACTTGGAAAAAACCTGA
AAGATGTTCATCAGGTTCTGAAAGAAGATATTGGCGATGCAATTGGCGGCGTGCACCTG
CTACCGTTTTTCCCGAGCACAGGGGATCGCGGCTTTGCGCCGGCGGACTATACACGCGT
TGATGCAGCGTTTGGCGACTGGGCAGACGTCGAGGCACTGGGCGAAGAGTACTACCTGA
TGTTTGATTTTATGATAAACCATATCTCCAGAGAGTCTGTGATGTATCAGGATTTTAAAA
AGAATCATGATGATTCAAAATATAAAGATTTCTTTATTCGTTGGGAAAAATTTTGGGCTA
AAGCTGGAGAAAATAGACCGACACAAGCAGATGTTGATCTGATTTATAAAAGAAAAGAT
AAAGCTCCGACACAAGAAATTACATTTGATGATGGAACAACAGAAAATTTATGGAATACA
TTTGGAGAAGAACAAATTGATATTGATGTTAATTCTGCTATTGCTAAGGAGTTCATCAAA
ACAACGCTCGAGGATATGGTCAAACATGGCGCGAACCTGATCCGCCTGGACGCGTTCGC
GTACGCCGTGAAAAAAGTTGACACAAACGACTTTTTCGTAGAACCTGAAATCTGGGATAC
ACTGAACGAAGTCCGCGAAATCCTGACCCCGCTGAAAGCGGAGATTCTGCCTGAAATTC
ATGAACATTACAGCATCCCTAAGAAGATCAATGATCATGGCTATTTCACGTACGATTTCG
CGCTTCCGATGACGACGCTTTATACACTGTATAGCGGCAAAACCAACCAGCTTGCTAAAT
GGTTAAAAATGTCTCCGATGAAGCAGTTCACGACCCTGGACACCCATGACGGGATCGGA
GTTGTCGACGCGCGCGACATTTTGACAGACGACGAGATTGATTATGCGTCTGAACAGCT
GTATAAGGTAGGCGCGAACGTCAAGAAGACGTATAGCAGCGCAAGCTACAACAACCTGG
ATATCTATCAGATCAACTCCACATATTATTCCGCGCTCGGCAACGACGACGCGGCCTATC
TGCTTTCACGCGTGTTTCAGGTGTTTGCGCCGGGCATTCCGCAGATCTATTATGTGGGG
CTGCTGGCCGGAGAGAACGATATTGCGCTGCTTGAATCCACCAAGGAAGGCCGCAACAT
CAACCGGCATTACTACACACGGGAAGAAGTTAAATCTGAAGTGAAACGGCCGGTGGTCG
CAAATCTCCTGAAACTTCTGAGCTGGCGCAACGAAAGTCCGGCCTTTGACCTGGCGGGA
TCAATCACGGTGGACACGCCGACGGATACGACGATTGTGGTGACGAGACAGGATGAAAA
TGGCCAAAATAAAGCGGTGCTGACAGCGGATGCGGCCAACAAAACATTTGAAATCGTGG
AAAACGGCCAAACAGTCATGAGCAGCGATAACCTGACACAGAACCATCATCATCATCATC
ACTAA

Protein sequence for cotB-<u>linker</u>-*Lm*SucP-His-tag

**MSKRRMKYHSNNEISYYNFLHSMKDKIVTVYRGGPESKKGKLTAVKSDYIALQAEKKIIYY
QLEHVKSITEDTNNSTTTIETEEMLDADDFHSLIGHLINQSVQFNQGGPESKKGRLVWLG
DDYAALNTNEDGVVYFNIHHIKSISKHEPDLKIEEQTPVGVLEADDLSEVFKSLTHKWVSI
NRGGPEAIEGILVDNADGHYTIVKNQEVLRIYPFHIKSISLGPKGSYKKEDQKNEQNQEDN
NDKDSNSFISSKSYSSSKSSKRSLKSSDDQSSKSGRSSRSKSSSKSSKRSLKSSDYQSSKSG
RSSRSKSSSKSSKRSLKSSDYQSSKSSKRSPRSSDYQSSRSPGYSSSIKSSGKQKEDYSYETI
VRTIDYHWKRKF**<u>GGGGGSGGGGGG</u>**MEIQNKAMLITYADSLGKNLKDVHQVLKEDIGDAIG
GVHLLPFFPSTGDRGFAPADYTRVDAAFGDWADVEALGEEYYLMFDFMINHISRESVMYQD
FKKNHDDSKYKDFFIRWEKFWAKAGENRPTQADVDLIYKRKDKAPTQEITFDDGTTENLWN
TFGEEQIDIDVNSAIAKEFIKTTLEDMVKHGANLIRLDAFAYAVKKVDTNDFFVEPEIWDTLNE
VREILTPLKAEILPEIHEHYSIPKKINDHGYFTYDFALPMTTLYTLYSGKTNQLAKWLKMSPMK
QFTTLDTHDGIGVVDARDILTDDEIDYASEQLYKVGANVKKTYSSASYNNLDIYQINSTYYSA**

Chapter 6. Supplementary material

LGNDDAAYLLSRVFQVFAPGIPQIYYVGLLAGENDIALLESTKEGRNINRHYYTREEVKSEVKR
PVVANLLKLLSWRNESPAFDLAGSITVDTPTDTTIVVTRQDENGQNKAVLTADAANKTFEIVE
NGQTVMSSDNLTQNHHHHHH*

DNA sequence for cotY-linker-*Ba*SucP (p02033)

ATGAGCTGCGGAAAAACCCATGGCCGGCATGAGAACTGTGTATGCGATGCAGTGGAAA
AGATTTTAGCAGAGCAGGAGGCAGTTGAAGAACAGTGTCCGACTGGCTGCTATACCAA
CCTTTTAAACCCTACGATTGCTGGAAAAGACACAATTCCGTTTCTCGTTTTTGATAAAAA
AGGCGGATTGTTCTCCACATTCGGAAACGTAGGGGGATTTGTGGATGATATGCAATGC
TTTGAATCCATTTTCTTCCGCGTCGAAAAATTATGCGATTGCTGTGCAACACTGTCTAT
TTTACGCCCGGTCGATGTCAAAGGCGATACCTTAAGTGTTTGCCACCCTTGCGACCCGG
ATTTCTTCGGGCTAGAAAAAACAGATTTCTGCATTGAAGTGGATCTCGGATGCTTCTGC
GCGATTCAGTGCCTGTCACCAGAGCTAGTTGACAGAACATCGCCTCACAAAGATAAAAA
GCATCATCACAATGGA<u>GGTGGAGGTGGTGGAAGTGGTGGAGGTGGTGGAGGT</u>ATGAAA
AACAAAGTTCAGCTGATTACATACGCCGACCGGCTTGGCGATGGGACAATTAAATCGAT
GACAGATATTCTTCGGACGCGCTTCGACGGAGTTTACGACGGAGTGCACATCCTTCCGT
TCTTTACACCTTTTGACGGTGCTGACGCCGGGTTTGATCCAATTGATCACACGAAAGTGG
ATGAGCGACTGGGCTCTTGGGATGATGTCGCCGAACTGTCCAAAACGCATAATATTATG
GTGGACGCCATCGTTAACCATATGTCTTGGGAAAGTAAACAATTTCAGGATGTGTTGGC
GAAAGGGGAGGAAAGCGAATACTATCCGATGTTTTTGACAATGTCAAGTGTATTTCCGA
ATGGGGCAACGGAAGAAGACCTTGCAGGAATCTACAGACCGCGGCCGGGACTGCCGTTT
ACTCACTATAAGTTTGCTGGCAAAACCAGACTTGTTTGGGTAAGCTTTACACCGCAGCAG
GTCGATATCGATACTGATAGCGACAAAGGATGGGAGTACTTAATGTCCATTTTTGATCAG
ATGGCGGCGTCACACGTCAGTTACATCCGACTTGACGCAGTCGGTTACGGTGCCAAAGA
GGCTGGCACAAGCTGCTTTATGACACCAAAAACGTTTAAGTTGATCTCTCGTCTGCGAGA
AGAAGGCGTCAAGCGAGGCTTGGAAATTCTGATTGAGGTACACTCATATTACAAAAAAC
AGGTTGAAATTGCCAGCAAAGTCGACAGGGTTTACGATTTTGCGCTTCCACCTCTCCTCC
TCCACGCCTTATCAACAGGACACGTGGAACCAGTAGCCCATTGGACGGATATTAGACCG
AATAATGCTGTTACTGTCCTGGATACCCACGATGGCATAGGGGTTATTGACATCGGCTCT
GATCAACTTGACCGTTCACTTAAAGGCCTTGTGCCGGACGAAGATGTTGACAATCTTGTA
AATACAATTCATGCCAATACACACGGAGAATCTGAGGCCGCGACTGGGGCAGCGGCAAG
TAACTTGGACCTGTATCAAGTTAACTCTACCTACTACTCCGCGTTAGGATGCAATGATCA
ACACTACATTGCGGCAAGAGCGGTCCAGTTTTTCCTGCCTGGGGTCCCTCAAGTCTACTA
CGTTGGAGCACTGGCAGGCAAGAATGATATGGAATTGTTGAATAAAACTAATAATGGCA
GAGATATAAATCGGCACTATTACTCGACGGCTGAAATCGACGAGAACTTAAAACGGCCG
GTCGTTAAAGCCCTTAATGCATTAGCAAAATTTAGAAATGAGCTCGACGCATTTGACGGG
ACATTCTCTTACACGACACCGACAGACACTTCTATTTCCTTCACTTGGAGAGGGGAAACA
TCGGAAGCAACCTTAACTTTTGAACCAAAACGCGGCTTGGGAGTCGATAATACGACTCCA
GTTGCAATGCTGGAATGGCATGATAGCGCGGGTGACCACAGATCTGATGATCTCATTGC
GAATCCGCCGGTTGTGGCGTAA

Protein sequence for cotY-<u>linker</u>-*BaSucP*

MSCGKTHGRHENCVCDAVEKILAEQEAVEEQCPTGCYTNLLNPTIAGKDTIPFLVFDKKG
GLFSTFGNVGGFVDDMQCFESIFFRVEKLCDCCATLSILRPVDVKGDTLSVCHPCDPDFFG
LEKTDFCIEVDLGCFCAIQCLSPELVDRTSPHKDKKHHHN<u>GGGGGSGGGGGG</u>MKNKVQ
LITYADRLGDGTIKSMTDILRTRFDGVYDGVHILPFFTPFDGADAGFDPIDHTKVDERLGSWD
DVAELSKTHNIMVDAIVNHMSWESKQFQDVLAKGEESEYYPMFLTMSSVFPNGATEEDLAG
IYRPRPGLPFTHYKFAGKTRLVWVSFTPQQVDIDTDSDKGWEYLMSIFDQMAASHVSYIRLD
AVGYGAKEAGTSCFMTPKTFKLISRLREEGVKRGLEILIEVHSYYKKQVEIASKVDRVYDFALP
PLLLHALSTGHVEPVAHWTDIRPNNAVTVLDTHDGIGVIDIGSDQLDRSLKGLVPDEDVDNL
VNTIHANTHGESEAATGAAASNLDLYQVNSTYYSALGCNDQHYIAARAVQFFLPGVPQVYY
VGALAGKNDMELLNKTNNGRDINRHYYSTAEIDENLKRPVVKALNALAKFRNELDAFDGTFS
YTTPTDTSISFTWRGETSEATLTFEPKRGLGVDNTTPVAMLEWHDSAGDHRSDDLIANPPVV
A*

6.7   Oligonucleotides and plasmid maps

S8.1 Oligonucleotides employed in Chapter 3.

Table 6.1: List of oligos employed for the construction of plasmids used in Chapter 3.
The bold sequence indicates the binding region of the oligos.

Part (bp)	Oligo ID	Sequence (5'→ 3')	Template
Plasmid p02002			
cwlD front (530 bp)	02024 cwlD front fw	AGATCTTCCGGATGGCTCGA GTTTTTCAGCAAGAT**CTTCAT TTGGGCGAATTTCC**	Bs02001
	02025 cwlD front rv	TACGAACGGTAGGCCTCGAG GATCCAATTCGTGATT**ACCGC ATGCGAAGCTTAC**	
cwlD back (551 bp)	02026 cwlD back fw	CTCTAGATGAATTGGTGAAGC GCTGAT**CAGAAAAAGGAGAC CCTCCGGAGTAATG**	Bs02001
	02027 cwlD back rv	AGCTGAGAATATTGTAGGAG ATCTTCTAGAAAGAT**CCGCAT CAATTAAACCAACC**	

Table 6.1: List of oligos employed for the construction of plasmids used in Chapter 3. The bold sequence indicates the binding region of the oligos (continued).

Part (bp)	Oligo ID	Sequence (5'→ 3')	Template
Plasmid p02005			
sleB front (556 bp)	02012 sleB front fw	AGATCTTCCGGATGGCTCGA GTTTTTCAGCAAGAT**CTGCAC ACAAAAAAGCCGC**	Bs02001
	02023 sleB front rv	TACGAACGGTAGGCCTCGAG GATCCAATTCGTGAT**CTTCGG GGTTTTTGGAGG**	
sleB back (541 bp)	02014 sleB back fw	CGGTAGGCCTCTAGATGAATT GGTGAAGCGCTGAT**CATTTCC CGGCTGAATTTGC**	Bs02001
	02022 sleB back rv	AGCTGAGAATATTGTAGGAG ATCTTCTAGAAAGAT**GCAGAA CCTATATATAAAGACC**	
Plasmid p02004			
spoIIGA front (440 bp)	02039 spoIIGA front fw	AGATCTTCCGGATGGCTCGA GTTTTTCAGCAAGAT**CCTATT GCTTCCTTCGCTTAG**	Bs02001
	02040 spoIIGA front rv	TACGAACGGTAGGCCTCGAG GATCCAATTCGT**GATCGAGG AAGTATAATGAGAGGATATA**	
spoIIGA back-spec (1,491 bp)	14089 Spec mid fw	**CAATAGCCAAATCAGGATCA TAGC**	p14028 (PhD book of F. Nadler)
	14014 pJET1.2 Seq rev	**AAGAACATCGATTTTCCATG GCAG**	

Table 6.1: List of oligos employed for the construction of plasmids used in Chapter 3. The bold sequence indicates the binding region of the oligos (continued).

Part (bp)	Oligo ID	Sequence (5'→ 3')	Template
Plasmid p02006			
cotB back (567 bp)	02049 cotB back SSS oH fw	CGGTAGGCCTCTAGATGAATT GGTGAAGCGCTGAT**GAGGAG CATCAGATAAAGTG**	Bs02001
	02050 cotB back oH rv	AGCTGAGAATATTGTAGGAG ATCTTCTAGAAAGAT**GTATAC GGTGTTTGTTCAGTTG**	
cotB front (589 bp)	02047 cotB front pJET OH fw	AGATCTTCCGGATGGCTCGA GTTTTTCAGCAAGAT**AGCTCA GTTTAAGTCAGAAGTG**	Bs02001
	02062 cotB front SSS oH fw	TACGAACGGTAGGCCTCGAG GATCCAATTCGTGAT**CGCGAA TACCCATTTTCACG**	
Plasmid p02019			
cotA front (569 bp)	02089 cotA frn OH fw	AGATCTTCCGGATGGCTCGA GTTTTTCAGCAAGAT**CCATGT TCATCTGTTCTTTTTGG**	Bs02001
	02090 cotA frn OH rv	TACGAACGGTAGGCCTCGAG GATCCAATTCGTGAT**CCTCTA GCAGGCAAGTTTTTTC**	
cotA back (582 bp)	02132 cotA back OH fw	AACAAGTGTGGATGCTTTCCT CTTATTTAGGGTAAA**CAAAAT CCTAAACGGCAGG**	Bs02001
	02092 cotA back OH rv	AGCTGAGAATATTGTAGGAG ATCTTCTAGAAAGAT**TTGCAG AAAGAAATGAAGCG**	

Table 6.1: List of oligos employed for the construction of plasmids used in Chapter 3. The bold sequence indicates the binding region of the oligos (continued).

Part (bp)	Oligo ID	Sequence (5'→ 3')	Template
mrgA (532 bp)	02130 mrgA OH rv	TCAAATCCTAAACGGCCTGCC GTTTAGGATTTTGT**TTACCCT AAATAAGAGGAAAGCATC**	Bs02001
	02129 mrgA OH fw	CTAAATAAAAAGATTAACTAA TAAGGAGGACAAAC**ATGAAA ACTGAAAACGCAAAAAC**	
sspB (202 bp)	02131 PsspB cotA OH fw	CGGTAGGCCTCTAGATGAATT GGTGAAGCGCTGAT**CTCCGC ATGATTTTCCGGCC**	Bs02001
	02128 PsspB Ro OH rv	GTTTGTCCTCCTTATTAGTTA ATC**TTTTTATTTAGTATGGTT GGG**	

Plasmid p02023

Part (bp)	Oligo ID	Sequence (5'→ 3')	Template
spo0E front (494 bp)	02158 spo0E front fw	**CCACTTTTCATATTGTGACAG TTC**	Bs02001
	02159 spo0E front rv	**GCTAAGAAATAGGAAACAAG TTTG**	
spo0E back (481 bp)	02160 spo0E back fw	CGGTAGGCCTCTAGATGAATT GGTGAAGCGCTGAT**ATAAAG AGAGAGCTTTTCGGAAG**	Bs02001
	02161 spo0E back rv	AGCTGAGAATATTGTAGGAG ATCTTCTAGAAAGAT**GCATTG CCATTCGATTTGC**	

Plasmid p02130

Part (bp)	Oligo ID	Sequence (5'→ 3')	Template
cotB (1211 bp)	02173 cotB OH Pcoty fw	AAGTCAAATACCCTATAAAGA AGGAGCTGAAATCA**ATGAGC AAGAGGAGAATGAAATATC**	Bs02001

Table 6.1: List of oligos employed for the construction of plasmids used in Chapter 3. The bold sequence indicates the binding region of the oligos (continued).

Part (bp)	Oligo ID	Sequence (5'→ 3')	Template
	02172 cotB OH linker rv	ACTTCCACCACCTCCACCACT TCCACCACCTCCAC**CAAATTT ACGTTTCCAGTGATAGTC**	
sfGFP (788 bp)	02067 sfGFP lox71 OH rv	AATAACTTCGTATAATGTATG CTATACGAACGGTA**TCATTTG TACAGTTCATCCATACC**	pYTK001
	02066 sfGFP linker OH fw	GGTGGAGGTGGTGGAAGTGG TGGAGGTGGTGGAGG**TATGC GTAAAGGCGAAGAG**	
PcotYZ (184 bp)	02124 PcotYZ pksx OH fw	TTATAAAAATTGCCCTCTCAT TTTTTTCTTGTTACC**CAAGCA TATGATGAATATATAGACG**	Bs02001
	02171 PcotYZ cotB OH rv	ATTTCATTATTTGAATGATAT TTCATTCTCCTCTTGCTCATT **GATTTCAGCTCCTTCTTTATA GG**	
pJK179 (BB) (4279 bp)	02021 pksX back	CATAACTTCGTATAGCATACA TTATACGAACGGTA**GATGAAT TGGTGAAGCGCTG**	pJK179
	31055 targ rv	**GTAACAAGAAAAAAATGAGA GGG**	
spec (1094 bp)	02017 specR lox	**CTACCGTTCGTATAATGTATG CTATACGAAGTTATGGTCAG CTTTATTGAACAGTAATTTAA G**	p10015

Table 6.1: List of oligos employed for the construction of plasmids used in Chapter 3. The bold sequence indicates the binding region of the oligos (continued).

Part (bp)	Oligo ID	Sequence (5'→ 3')	Template
	02016 specR lox	**TACCGTTCGTATAGCATACAT TATACGAAGTTATTAGACCG GTACTAAATTAAAGTAATAAA G**	
Common oligos & parts for plasmids p02002-23			
spec (part 1) (1244 bp)	14249 SSS-lox71 fw	**ATCACGAATTGGATCCTCGA G**	BsFLN039 v.I. (PhD book of F. Nadler)
	14090 Spec-mid-rv	**GGACAAATTCAGGAACCAAG C**	
spec (part 2) (1078 bp)	14089 Spec-mid-fw	**CAATAGCCAAATCAGGATCA TAGC**	BsFLN039 v.I. (PhD book of F. Nadler)
	14250 SSS-lox66 rv	**ATCAGCGCTTCACCAATTCAT C**	
pJET (2951 bp)	23043 pJET fw	**GCTCGAGTTTTTCAGCAAGA T**	pJET vector (Thermo Fisher scientific)
	23044 pJET rv	**ATCTTTCTAGAAGATCTCC**	

S8.2 Oligonucleotides employed in Chapter 4.

Table 6.2: List of oligos employed for the construction of plasmids used in Chapter 4. The bold sequence indicates the primer binding region.

Part (bp)	Oligo ID	Sequence (5'→ 3')	Template
Plasmid p02012			
cotZ front (600 bp)	02082 cotZ fr fw	AGATCTTCCGGATGGCTCGA GTTTTTCAGCAAGAT**CAGCCG AACCAGTGAGTC**	Bs02001
	02088 cotZ fr rv	TACGAACGGTAGGCCTCGAG GATCCAATTCGTGATCCTGCA GGCTTTTCTCTATG	
cotZ back (561 bp)	02084 cotZ back fw	CGGTAGGCCTCTAGATGAATT GGTGAAGCGCTGAT**CCTGCC TTATAAATTTGAGACTG**	Bs02001
	02085 cotZ back rv	AGCTGAGAATATTGTAGGAG ATCTTCTAGAAAGATC**CTTTT AAACCCTACGATTGC**	
Plasmid p02024			
cotY front (512 bp)	02166 cotY fr fw	AGATCTTCCGGATGGCTCGA GTTTTTCAGCAAGATA**GTTGA TTGCGCAGAAGC**	Bs02001
	02167 cotY fr rv	TACGAACGGTAGGCCTCGAG GATCCAATTCGTGAT**CTTCCA AACTCTATTACAGGTGC**	
cotY back (636 bp)	02164 cotY bck fw	CGGTAGGCCTCTAGATGAATT GGTGAAGCGCTGAT**TGATTTC AGCTCCTTCTTTATAGG**	Bs02001
	02165 cotY bck fw	AGCTGAGAATATTGTAGGAG ATCTTCTAGAAAGATA**TCCTG ACTGTGACCATCC**	

Table 6.2: List of oligos employed for the construction of plasmids used in Chapter 4. The bold sequence indicates the primer binding region (continued).

Part (bp)	Oligo ID	Sequence (5'→ 3')	Template
Plasmid p02022			
*Lm*SucP (1527 bp)	02154 sucP OH fw	GGTGGAGGTGGTGGAAGTGG TGGAGGTGGTGGAGGT**ATGG AAATTCAAAATAAAGCGATG**	p02017
	02144 sucP His tag rv	TTAGTGATGATGATGATGATG **GTTCTGTGTCAGGTTATCGC**	
cotZ (444 bp)	02145 cotZ fw	**ATGAGCCAGAAAACATCAAG C**	Bs02001
	02147 cotZ rv	**ATGATGATGTGTACGATTGA TTAATC**	
PcotYZ-pJK179 (4351 bp)	10152 pksX-back fw	**ATGAATTGGTGAAGCGCTGA G**	p17001 (BSc book of D. Klewinghaus)
	02146 PcotYZ rv	**TGATTTCAGCTCCTTCTTTAT AGG**	
spec (1094 bp)	02017 specR lox	**CTACCGTTCGTATAATGTATG CTATACGAAGTTATGGTCAG CTTTATTGAACAGTAATTTAA G**	p10015
	02016 specR lox	**TACCGTTCGTATAGCATACAT TATACGAAGTTATTAGACCG GTACTAAATTAAAGTAATAAA G**	
Plasmid p02025			

Chapter 6. Supplementary material

Table 6.2: List of oligos employed for the construction of plasmids used in Chapter 4. The bold sequence indicates the primer binding region (continued).

Part (bp)	Oligo ID	Sequence (5'→ 3')	Template
*Lm*SucP (1527 bp)	02154 sucP OH fw	GGTGGAGGTGGTGGAAGTGG TGGAGGTGGTGGAGGT**ATGG AAATTCAAAATAAAGCGATG**	p02017
	02144 sucP His tag rv	TTAGTGATGATGATGATGATG **GTTCTGTGTCAGGTTATCGC**	
cotB-pJK179 (6652 bp)	02172 cotB OH linker rv	ACTTCCACCACCTCCACCACT TCCACCACCTCCACC**AAATTT ACGTTTCCAGTGATAGTC**	
	o02003 sucP spec	GACACAGAACCATCATCATCA TCATCACTAAT**ACCGTTCGTA TAGCATACATTATACGAAGTT ATTAGACCG**	
Plasmid p02027			
CvFAP (part1) (955 bp)	02175 cvFAP OH linker fw	GGTGGAGGTGGTGGAAGTGG TGGAGGTGGTGGAAGT**ATGA GAGCGAGCGCAG**	p02010
	02178 cvFAP seq rv	**TTCCAAATTCTTTCAGTTCTGC**	
CvFAP (part2) (1056 bp)	02176 cvFAP OH lox71 rv	CGTATAATGTATGCTATACGA ACGGTACGCGTATATTAGTGA TGATGATGATGATG**TGCTGC AACTGTCGCC**	p02010
	02177 cvFAP seq fw	**ACCGACAGGAGAAAGACTG**	

Table 6.2: List of oligos employed for the construction of plasmids used in Chapter 4. The bold sequence indicates the primer binding region (continued).

Part (bp)	Oligo ID	Sequence (5'→ 3')	Template
cotY-pJK179 (5067 bp)	02016 specR lox	**TACCGTTCGTATAGCATACAT TATACGAAGTTATTAGACCG GTACTAAATTAAAGTAATAAA G**	p02026
	02174 cotY OH rv	ACTTCCACCACCTCCACCACT TCCACCACCTCCACC**TCCATT GTGATGATGCTTTTTATC**	
Plasmid p02030			
cotY-CvFAP (part1) (1584 bp)	02221 cotY pBReT fw	AAGCGTTGGTTTGGCAATCTT ATCGGGCTATGCAT**CCAAGC ATATGATGAATATATAGACG**	Bs02035
	02178 cvFAP seq rv	**TTCCAAATTCTTTCAGTTCTG C**	
cotY-CvFAP (part2) (1126 bp)	02222 pksX pBreT rv	CGCTCAGTGGAACGAGGCGC CTGATCAACGCGTGC**GGCGT TTGTGTTTTCTCAG**	
	02177 cvFAP seq fw	**ACCGACAGGAGAAAGACTG**	
pMSE3 (BB) (5321 bp)	02220 pBreT fw	**GCACGCGTTGATCAGGC**	p14037 (PhD book of F. Nadler)
	02219 pBreT rv	**ATGCATAGCCCGATAAGATT G**	
Common oligos for plasmids p02012-24			

Table 6.2: List of oligos employed for the construction of plasmids used in Chapter 4. The bold sequence indicates the primer binding region (continued).

Part (bp)	Oligo ID	Sequence (5'→ 3')	Template
spec (part 1) (1244 bp)	14249 SSS-lox71 fw	**ATCACGAATTGGATCCTCGAG**	BsFLN039 v.I. (PhD book of F. Nadler)
	14090 Spec-mid-rv	**GGACAAATTCAGGAACCAAGC**	
spec (part 2) (1078 bp)	14089 Spec-mid-fw	**CAATAGCCAAATCAGGATCATAGC**	BsFLN039 v.I. (PhD book of F. Nadler)
	14250 SSS-lox66 rv	**ATCAGCGCTTCACCAATTCATC**	
pJET (2951 bp)	23043 pJET fw	**GCTCGAGTTTTTCAGCAAGAT**	pJET vector (Thermo Fisher scientific)
	23044 pJET rv	**ATCTTTCTAGAAGATCTCC**	

S8.3 Assembly of plasmid p02022 using LCR

Table 6.3: List of bridging oligos employed for the assembly of p02022. Each bridging oligo assembles two neighboring parts.

	Parts	Bridging oligo ID	Sequence (5'→ 3')
1.	*PcotYZ-pJK179*	o02008 PcoYZ cotZ	**AGTCAAATACCCTATAAAGAAGG AGCTGAAATCAATGAGCCAGAA AACATCAAGCTGCG**
2.	*cotZ*		
1.	*cotZ*	o02009 cotZ linker	**CCTCGATTAATCAATCGTACACA TCATCATGGTGGAGGTGGTGGA AGTGGTG**
2.	*LmSucP*		
1.	*LmSucP*	o02003 sucP spec	**GACACAGAACCATCATCATCATC ATCACTAATACCGTTCGTATAGC ATACATTATACGAAGTTATTAGA CCG**
2.	*spec*		
1.	*spec*	10159 LCR specR-lox66 pksXback fw	**ACCATAACTTCGTATAGCATACA TTATACGAACGGTAGATGAATTG GTGAAGCGCTGAGAAAACAC**
2.	*PcotYZ-pJK179*		

S8.4 Vector maps of all employed plasmids

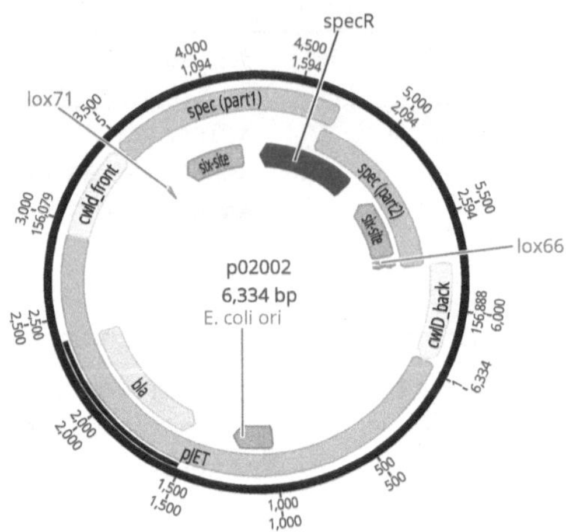

Figure S15: Vector map of p02002 integrative plasmid designed in this study for the deletion of *cwlD* in *B. subtilis* (Bs02004). **Annotations:** *bla*: gene for Ampicillin resistance, *lox71-lox66*: Cre recombinase recognition sites, *specR*: Spectinomycine resistance flanked by six-sites, *back-front* regions: homology regions upstream and downstream of the indicated locus used for genomic integration, pJET: backbone. All fragments were PCR amplified using oligos listed in Table 6.1. **Reference:** [86]

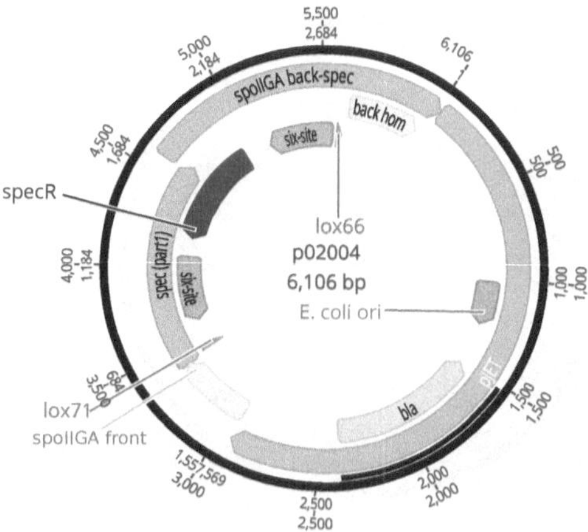

Figure S16: Vector map of p02004 integrative plasmid designed in this study for the deletion of *spoIIGA* in *B. subtilis* (Bs02005). **Annotations:** *bla*: gene for Ampicillin resistance, *lox71-lox66*: Cre recombinase recognition sites, *specR*: Spectinomycine resistance flanked by six-sites, *back-front* regions: homology regions upstream and downstream of the indicated locus used for genomic integration, pJET: backbone. All fragments were PCR amplified using oligos listed in Table 6.1. **Reference:** [86]

Chapter 6.   Supplementary material

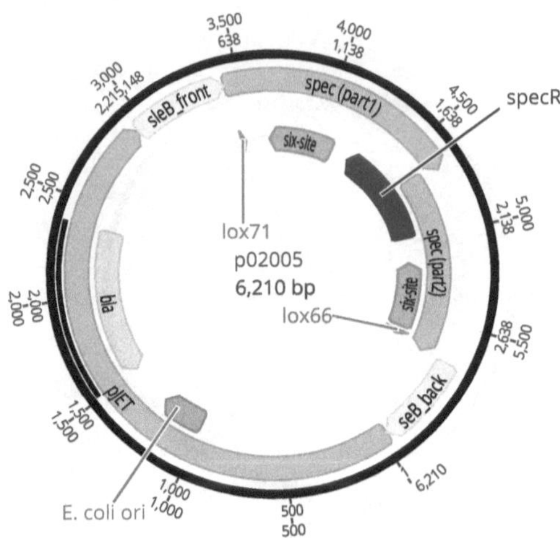

Figure S17: Vector map of p02005 integrative plasmid designed in this study for the deletion of *sleB* in *B. subtilis* (Bs02003). **Annotations:** *bla*: gene for Ampicillin resistance, *lox71-lox66*: Cre recombinase recognition sites, *specR*: Spectinomycine resistance flanked by six-sites, *back-front* regions: homology regions upstream and downstream of the indicated locus used for genomic integration, pJET: backbone. All fragments were PCR amplified using oligos listed in Table 6.1. **Reference:** [86]

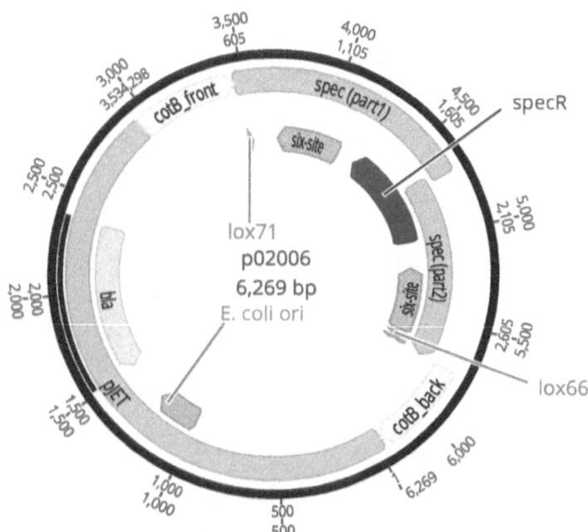

Figure S18: Vector map of p02006 integrative plasmid designed in this study for the deletion of *cotB* in *B. subtilis* (Bs02006). **Annotations:** *bla*: gene for Ampicillin resistance, *lox71-lox66*: Cre recombinase recognition sites, *specR*: Spectinomycine resistance flanked by six-sites, *back-front* regions: homology regions upstream and downstream of the indicated locus used for genomic integration of the spectinomycine resistance, pJET: backbone. All fragments were PCR amplified using oligos listed in Table 6.1.

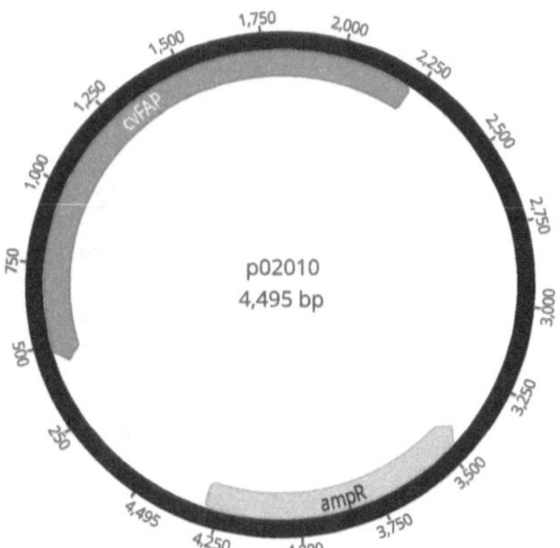

Figure S19: Vector map of p02010 plasmid used in this study as PCR template for the amplification of *CvFAP.* **Annotations:** *ampR:* Ampicillin resistance gene. **Reference:** Base-Clear B.V.

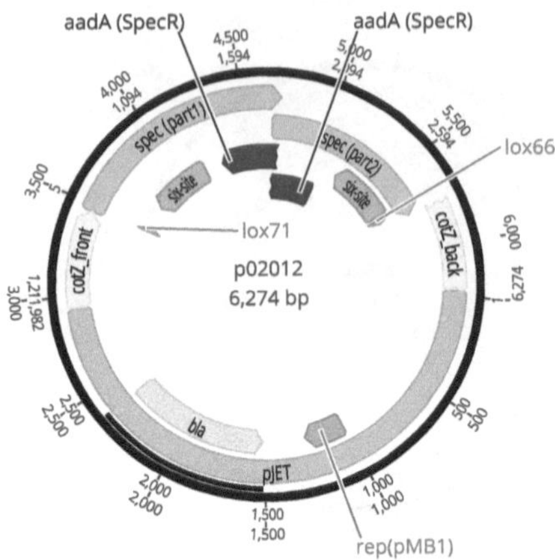

Figure S20: Vector map of p02012 integrative plasmid designed in this study for the deletion of *cotZ* in *B. subtilis*. **Annotations:** *bla*: gene for Ampicillin resistance, *lox71-lox66*: Cre recombinase recognition sites, *specR*: Spectinomycine resistance flanked by six-sites, *back-front* regions: homology regions upstream and downstream of the indicated locus used for genomic integration of the spectinomycine resistance, pJET: backbone. All fragments were PCR amplified using oligos listed in Table 6.2.

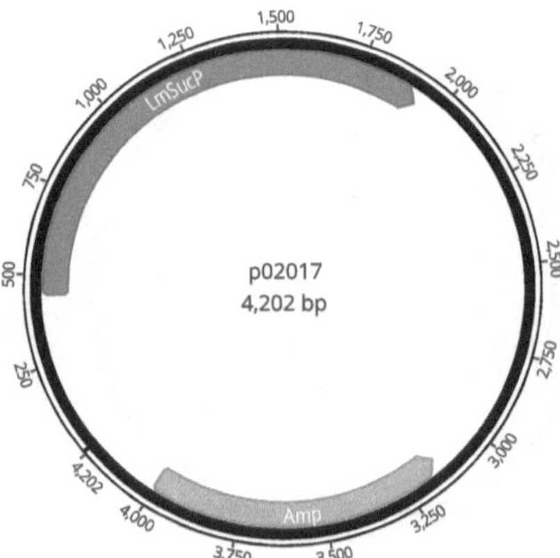

Figure S21: Vector map of p02017 plasmid used in this study as PCR template for the amplification of *Lm*SucP. **Annotations:** *amp:* Ampicillin resistance gene. **Reference:** Base-Clear B.V.

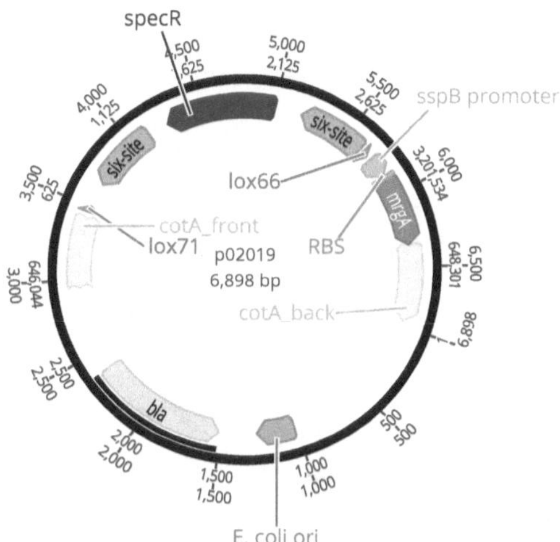

Figure S22: Vector map of p02019 integrative plasmid designed in this study for the deletion of *cotA* in *B. subtilis*. **Annotations:** *bla*: gene for Ampicillin resistance, *lox71-lox66*: Cre recombinase recognition sites, *specR*: Spectinomycine resistance flanked by six-sites, *back-front* regions: homology regions upstream and downstream of the indicated locus used for genomic integration of the spectinomycine resistance, pJET: backbone. All fragments were PCR amplified using oligos listed in Table 6.1. **Reference:** [86]

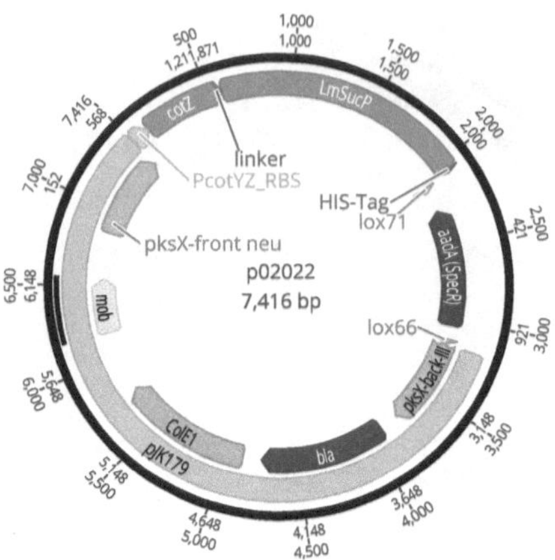

Figure S23: Vector map of p02022 integrative plasmid designed in this study for the display of *Lm*SucP as C-terminal fusion to the coat protein CotZ. Integration of the plasmid was done in the *pksX* locus of *B. subtilis*. **Annotations:** P*cotYZ*: *cotYZ* promoter, *bla*: gene for Ampicillin resistance, *lox71-lox66*: Cre recombinase recognition sites, *specR*: Spectinomycine resistance, *back-front* regions: homology regions upstream and downstream of the indicated locus used for genomic integration , ColE1: *E. coli* origin of replication, pJK179: backbone. All fragments were PCR amplified using oligos listed in Table 6.2.

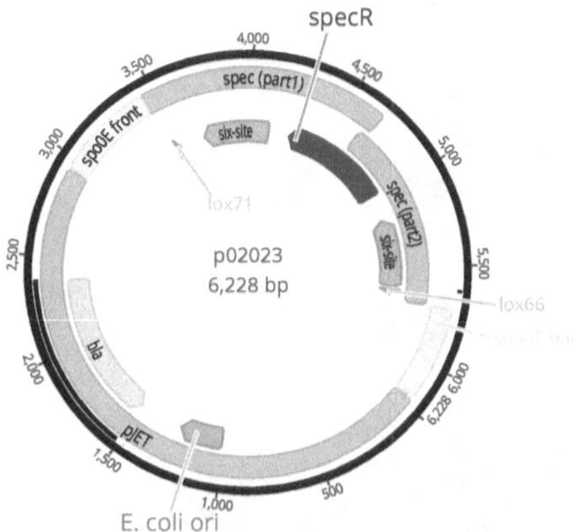

Figure S24: Vector map of p02023 integrative plasmid designed in this study for the deletion of *spoOE* in *B. subtilis* (Bs02025). **Annotations:** *bla*: gene for Ampicillin resistance, *lox71-lox66*: Cre recombinase recognition sites, *specR*: Spectinomycine resistance flanked by six-sites, *back-front* regions: homology regions upstream and downstream of the indicated locus used for genomic integration of the spectinomycine resistance, pJET: backbone. All fragments were PCR amplified using oligos listed in Table 6.1. **Reference:** [86]

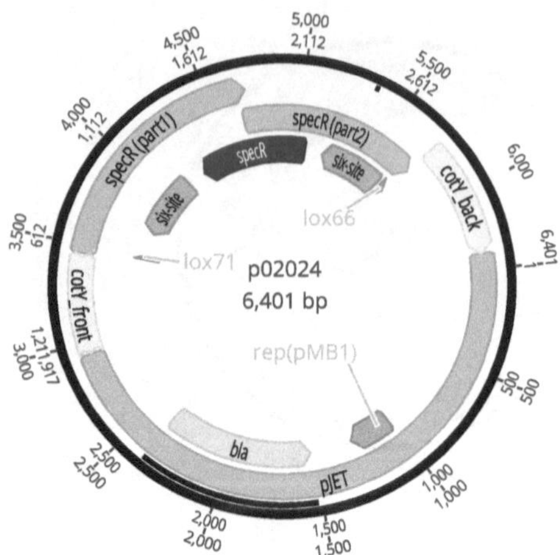

Figure S25: Vector map of p02024 integrative plasmid designed in this study for the deletion of *cotY* in *B. subtilis*. **Annotations:** *bla*: gene for Ampicillin resistance, *lox71-lox66*: Cre recombinase recognition sites, *specR*: Spectinomycine resistance flanked by six-sites, *back-front* regions: homology regions upstream and downstream of the indicated locus used for genomic integration of the spectinomycine resistance, pJET: backbone. All fragments were PCR amplified using oligos listed in Table 6.2.

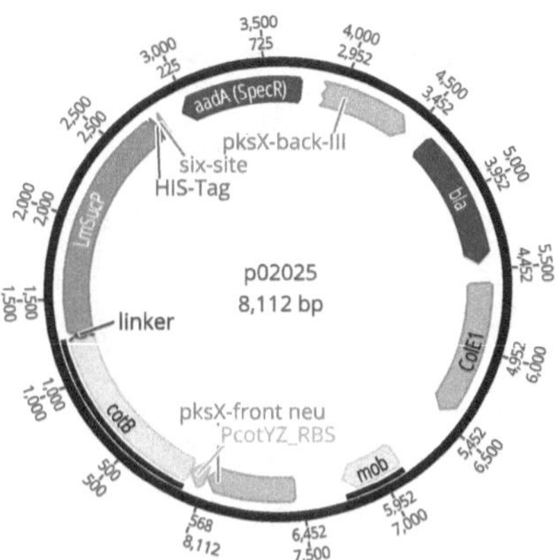

Figure S26: Vector map of p02025 integrative plasmid designed in this study for the display of *Lm*SucP as C-terminal fusion to the coat protein CotB. Integration of the plasmid was done in the *pksX* locus of *B. subtilis*. **Annotations:** P*cotYZ*: *cotYZ* promoter, *bla*: gene for Ampicillin resistance, *specR*: Spectinomycine resistance, *back-front* regions: homology regions upstream and downstream of the indicated locus used for genomic integration , ColE1: *E. coli* origin of replication, pJK179: backbone. All fragments were PCR amplified using oligos listed in Table 6.2.

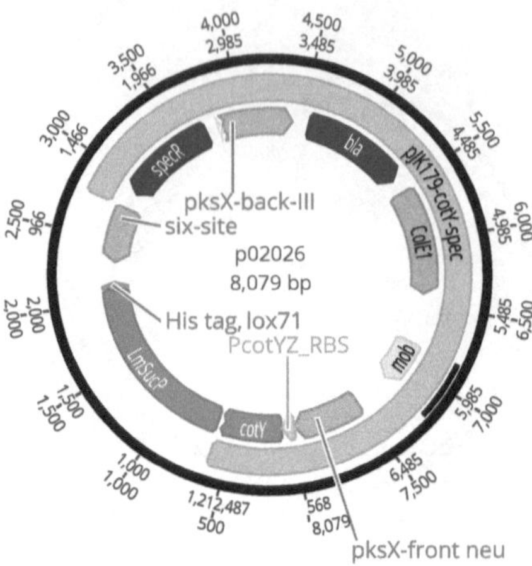

Figure S27: Vector map of p02026 used in this study as PCR template for the amplification of pJK179 backbone including the *pksX* back and front regions, *cotY* and *spec*. **Annotations:** P*cotYZ*: *cotYZ* promoter, *bla*: gene for Ampicillin resistance, *specR*: Spectinomycine resistance, *back-front* regions: homology regions upstream and downstream of the indicated locus used for genomic integration , ColE1: *E. coli* origin of replication, pJK179: backbone. All fragments were PCR amplified using oligos listed in Table 6.2.

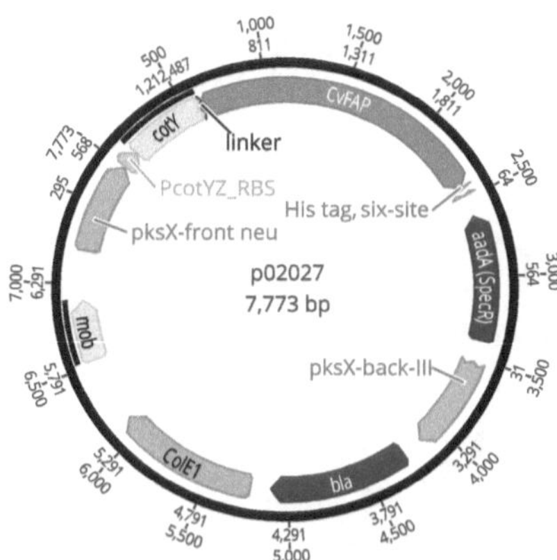

Figure S28: Vector map of p02027 integrative plasmid designed in this study for the display of *Cv*FAP as C-terminal fusion to the coat protein CotY. Integration of the plasmid was done in the *pksX* locus of *B. subtilis*. **Annotations:** *PcotYZ*: *cotYZ* promoter, *bla*: gene for Ampicillin resistance, *specR*: Spectinomycine resistance, *back-front* regions: homology regions upstream and downstream of the indicated locus used for genomic integration , ColE1: *E. coli* origin of replication, pJK179: backbone. All fragments were PCR amplified using oligos listed in Table 6.2.

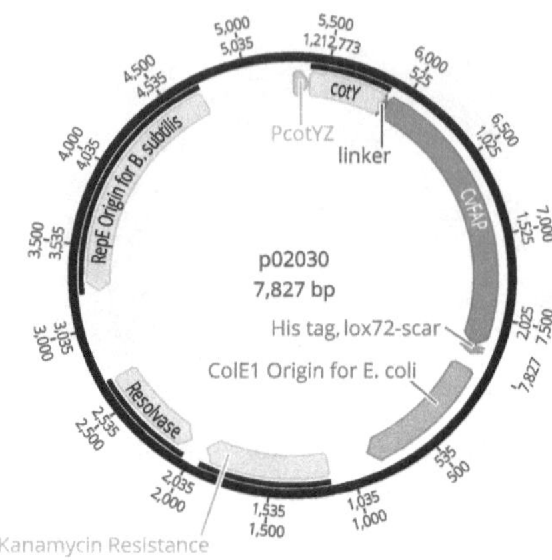

Figure S29: Vector map of p02030 high copy plasmid designed in this study for the display of *Cv*FAP as C-terminal fusion to the coat protein CotY. **Annotations:** P*cotYZ*: *cotYZ* promoter. All fragments were PCR amplified using oligos listed in Table 6.2.

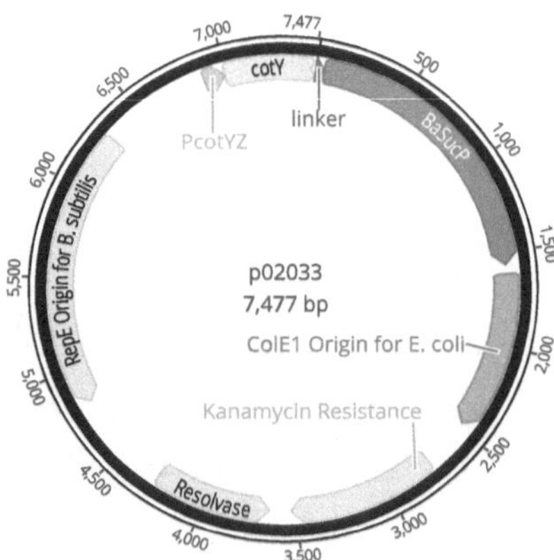

Figure S30: Vector map of p02033 high copy plasmid designed in this study for the display of *Ba*SucP as C-terminal fusion to the coat protein CotY. **Annotations:** P*cotYZ*: *cotYZ* promoter. **Reference:** TWIST Bioscience

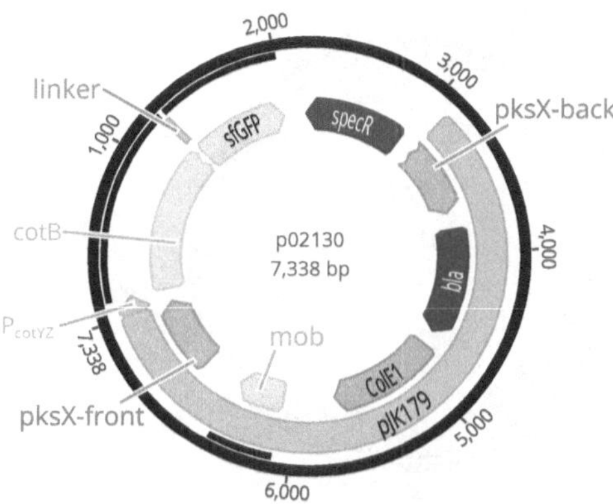

Figure S31: Vector map of p02130 integrative plasmid designed in this study for the display of sfGFP as a fusion to the CotB (C-terminal fusion). The construct is integrated in the *pksX* locus via homologous recombination. **Annotations:** *bla*: gene for Ampicillin resistance, *lox71-lox66*: Cre recombinase recognition sites, *specR*: Spectinomycine resistance flanked by six-sites, *back-front* regions: homology regions upstream and downstream of the indicated locus used for genomic integration, PcotYZ: promoter of *cotYZ*, ColE1: *E. coli* ori, pJET: backbone. All fragments were PCR amplified using oligos listed in Table 6.1. **Reference:** [86]

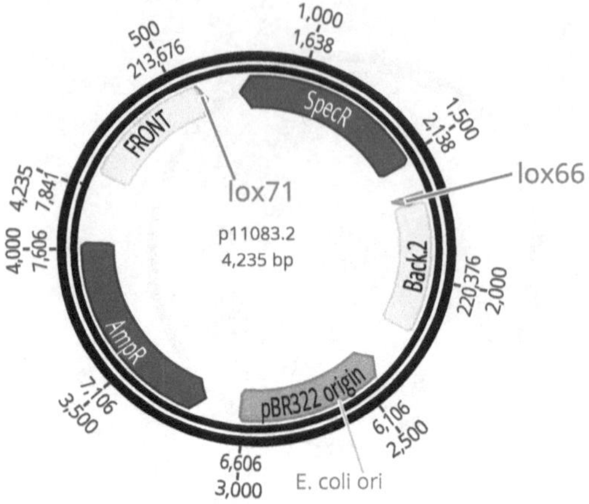

Figure S32: Vector map of p11083 integrative plasmid used in this study for the deletion of *skf* in *B. subtilis* (Bs02025). **Annotations:** *AmpR*: gene for Ampicillin resistance, *specR*: Spectinomycine resistance, *back-front* regions: homology regions upstream and downstream of the indicated locus used for genomic integration of the spectinomycine resistance. **Reference:** PhD book of S. Hackenschmidt.

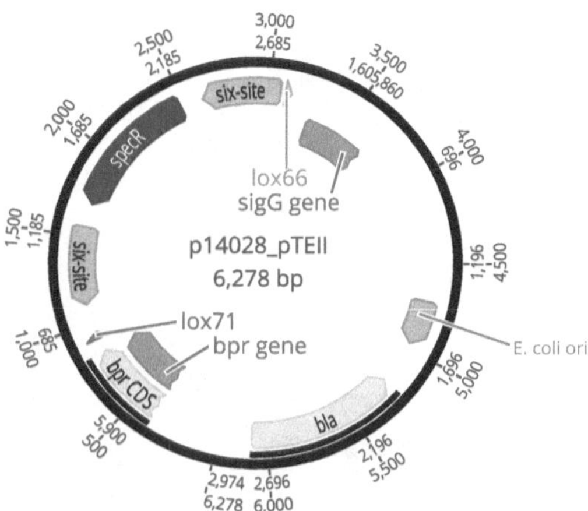

Figure S33: Vector map of p14028 integrative plasmid used in this book as PCR tem-plate for the amplification of *spoIIGA back-spec* for the construction of plasmid p02004. **Annotations:** *AmpR*: gene for Ampicillin resistance, *specR*: Spectinomycine resistance, *back-front* regions: homology regions upstream and downstream of the indicated locus used for genomic integration of the spectinomycine resistance. **Reference:** PhD book of F. Nadler.

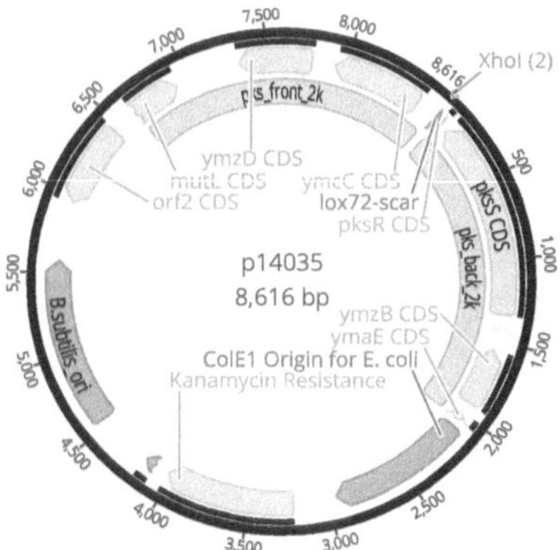

Figure S34: Vector map of p14035 low copy plasmid with two homology regions to the *pksX* locus. **Reference:** [131]

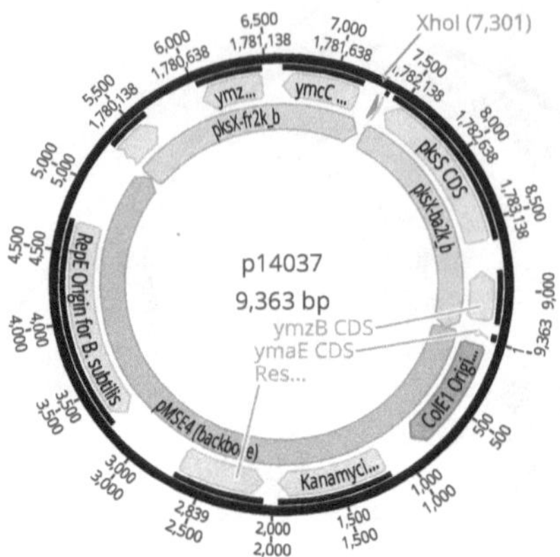

Figure S35: Vector map of p14037 high copy plasmid used as PCR template for the amplification of pMSE4 backbone. **Reference:** PhD book of F. Nadler.

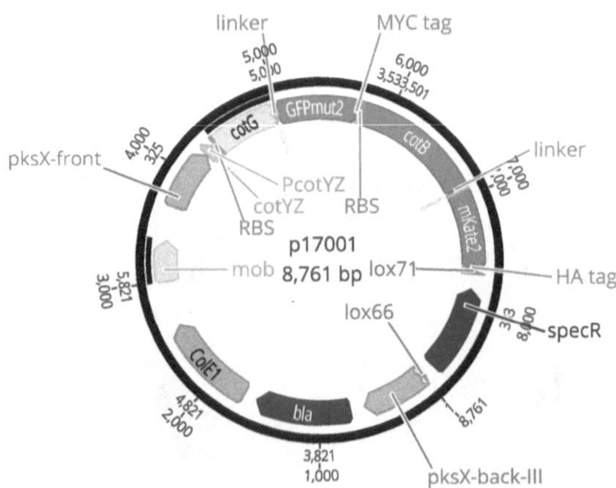

Figure S36: Vector map of p17001 plasmid used in this study as PCR template for the amplification of *PcotYZ*-pJK179 backbone for the construnction of plasmid p02022. **Annotations:** *bla*: gene for Ampicillin resistance, *specR*: Spectinomycine resistance, *pksX back-front* regions: homology regions upstream and downstream of the indicated locus, ColE1: *E. coli* ori. **Reference:** BSc book of D. Klewinghaus.

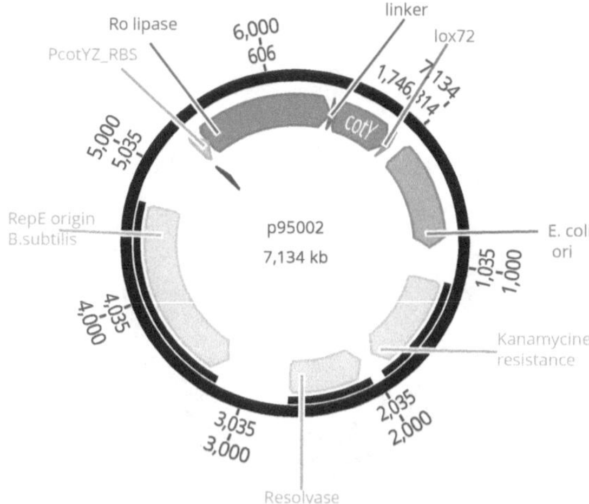

Figure S37: Vector map of p95002 high copy plasmid used for the spore display of RO lipase as a fusion to the CotY protein. **Reference:** BSc book of P. Gockel.

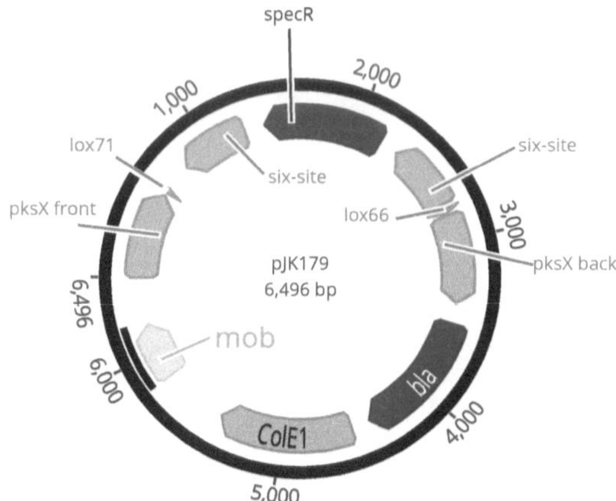

Figure S38: Vector map of pJK179 plasmid used in this study as PCR template for the amplification of pJK179 backbone including the *pksX* back and front regions. **Annotations:** *bla*: gene for Ampicillin resistance, *lox71-lox66*: Cre recombinase recognition sites, *specR*: Spectinomycine resistance flanked by six-sites, *back-front* regions: homology regions upstream and downstream of the indicated locus used for genomic integration, ColE1: *E. coli* ori. **Reference:** PhD book of J. Kabisch.

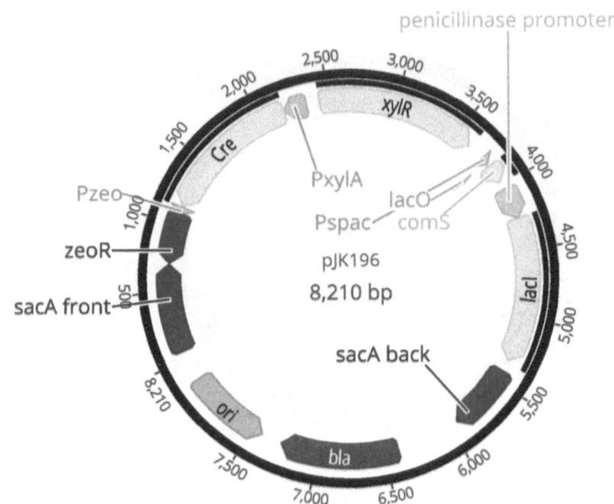

Figure S39: Vector map of pJK196 integrative plasmid used in this study for the integration of *comS* and *cre* (P*xylA*) operon into *sacA* locus of *B. subtilis* (Bs02002) with Zeo resistance. **Annotations:** *bla*: gene for Ampicillin resistance, *zeoR*: Zeocine resistance gene, P*xylA*: xylose inducible promoter, P*spac*: IPTG inducible promoter, *lacI*: lactose (lac) operon repressor, *lacO*: lac operator, xylR: xylose operon repressor *back-front* regions: homology regions upstream and downstream of the indicated locus used for genomic integration. **Reference:** [210]

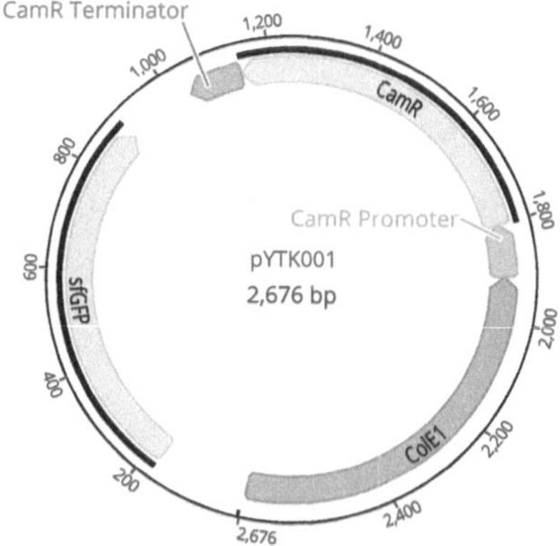

Figure S40: Vector map of pYTK001 integrative plasmid used in this study as PCR template for amplification of *sfgfp*. **Annotations:** *CamR*: chloramphenicol resistance gene, ColE1: *E. coli* origin of replication. **Reference:** [106]